T0224970

SpringerBriefs in Applied Sciences and Technology

Computational Intelligence

Series Editor

Janusz Kacprzyk, Systems Research Institute, Polish Academy of Sciences, Warsaw, Poland

SpringerBriefs in Computational Intelligence are a series of slim high-quality publications encompassing the entire spectrum of Computational Intelligence. Featuring compact volumes of 50 to 125 pages (approximately 20,000–45,000 words), Briefs are shorter than a conventional book but longer than a journal article. Thus Briefs serve as timely, concise tools for students, researchers, and professionals.

Jagdish Chand Bansal · Prathu Bajpai ·
Anjali Rawat · Atulya K. Nagar

Sine Cosine Algorithm for Optimization

 Springer

Jagdish Chand Bansal
Department of Mathematics
South Asian University
New Delhi, Delhi, India

Prathu Bajpai
Department of Mathematics
South Asian University
New Delhi, Delhi, India

Anjali Rawat
Department of Mathematics
National Institute of Technology
Aizawl, Mizoram, India

Atulya K. Nagar
School of Mathematics, Computer Science
and Engineering
Liverpool Hope University
Liverpool, UK

ISSN 2191-530X   ISSN 2191-5318 (electronic)
SpringerBriefs in Applied Sciences and Technology
ISSN 2625-3704   ISSN 2625-3712 (electronic)
SpringerBriefs in Computational Intelligence
ISBN 978-981-19-9721-1   ISBN 978-981-19-9722-8 (eBook)
https://doi.org/10.1007/978-981-19-9722-8

This Springer imprint is published by the registered company Springer Nature Singapore Pte Ltd.
The registered company address is: 152 Beach Road, #21-01/04 Gateway East, Singapore 189721, Singapore

Foreword

How to solve real-world complex optimization problems is important in applied systems analysis. In such applications, one often wants an algorithm which is easy to implement, uses fewer parameters, and has an efficient optimization capability; all of this makes sine cosine algorithm a good candidate for such scenarios.

The area of meta-heuristics has an abundance of swarm intelligence and evolutionary algorithms in the literature. As a population-based meta-heuristic sine cosine algorithm, although relatively nascent, is becoming popular in the research community. This monograph is an outcome of the constant efforts from the side of Dr. Jagdish Chand Bansal and Professor Atulya K. Nagar. The idea behind this book is to present a very focused volume discussing the sine cosine algorithm in a variety of optimization problems, like single objective optimization problems, multi-objective optimization problems, and combinatorial or discrete optimization problems. This book attempts to capture the attention of the research communities, engaged in diverse disciplines, who wish to incorporate the robust and dynamic nature of the meta-heuristic algorithms and are ready to celebrate the inherent randomness of these modern optimization tools.

I have many years of academic association, and joint publications, with Professor Nagar, and our work has always been on novel as well as well-focused research themes. Continuing in the same vein, it has been a real delight as I embarked on the pleasant journey of reading through this interesting book and have been convinced that this is one of the niche sources of introductory material on this topic to be found. The book's title, "Sine Cosine Algorithm for Optimization," aptly describes its aim and objectives, and the authors have achieved their motivation remarkably successfully providing a good coverage of theory and applications. The chapters are exemplary in giving useful insights for applications of the framework covered. Those of us who have devoted a substantial portion of our academic life and energies to the study and elaboration of optimization methods and mathematical sciences often wish we had a simple way of communicating and passing along the core concepts to newcomers and researchers to the topic area. I highly recommend this book for students and researchers who want to get into the basics of sine cosine optimization technique and who have an interest in equipping themselves to probe more deeply into this topic for a variety of application interests. If my reckoning is not completely amiss, those who read this monograph will find abundant reasons for sharing my conviction that we owe its authors a true debt of gratitude for putting this work together.

Prof. Kumbakonam Govindarajan
Subramanian
Formerly Professor at the Department
of Mathematics
Madras Christian College
Chennai, India

Distinguished Senior Visiting Professor
at the School of Mathematics
Computer Science and Engineering at
Liverpool Hope University, Hope Park
Liverpool, UK

Preface

Meta-heuristic algorithms are gaining popularity in various disciplines of science and engineering. In recent years, the need for robust optimization algorithms has drawn significant attention from the research community in developing new intelligent optimization techniques. Sine cosine algorithm is a relatively new algorithm in the field of meta-heuristic algorithms. Easy implementation, fewer parameters, and efficient optimization capabilities make sine cosine algorithm a good candidate for solving real-world complex optimization problems. Therefore, this book presents the latest developments in the sine cosine algorithm. The book emphasizes introducing sine cosine algorithm in the arena of the single-objective optimization problem, multi-objective optimization problems, and discrete optimization problems. The book also discusses the practical applications of the sine cosine algorithm. It includes recent research by various researchers and authors to give an overview of the latest developments to the readers. The book's content follows a logical order, and utmost care has been taken to make this book appealing to readers of various disciplines of science and engineering. This book is intended to serve as an important reference for postgraduate level students and researchers who wish to utilize sine cosine algorithm as a tool for optimization in their academic and research work.

New Delhi, India   Jagdish Chand Bansal
New Delhi, India   Prathu Bajpai
Aizawl, India   Anjali Rawat
Liverpool, UK   Atulya K. Nagar
October 2022

Acknowledgements The idea of presenting this particular book would not be possible without the financial support of Liverpool Hope University, Liverpool, UK. We would like to thank our publisher Springer Nature for constantly motivating us for writing this volume. We acknowledge the support of Shitu Singh, A. M. Mohiuddin, and Probhat Pouchang for their useful comments and discussions to present this book in a more reader friendly manner.

Contents

Chapter 1
Introduction

Decision-making is a difficult task, and it requires careful analysis of the underlying problem at hand. The presence of various alternative solutions makes the decision-making problem even more difficult as all the available solutions are not optimal. Since resources, time, and money are limited, or even sometimes scarce, the quest for optimal choices is of paramount importance for the welfare of the mankind. Optimization is a mathematical tool and an indispensable part of the decision-making process which assists in finding optimal (or near optimal) solutions from the set of available solutions. Optimization as a subject spans over almost every field of science and engineering and is mainly concerned with planning and design problems. For instance, in industrial design, corporate planning, budget planning, or holiday planning, optimization plays an important part in the decision-making. The need of optimization as a technique cannot be separated from different fields, such as computer science, engineering, medicine science, economics, and many more others disciplines. Advancements in the computational capabilities and availability of high-speed processors in modern computers have made optimization techniques more user friendly to tackle real-world optimization problems. In addition, easy access to advanced computer simulation techniques has prompted researchers to look for more generalized optimization methods which involve high computations, and are capable of handling more complex real-world optimization problems.

A general optimization problem can be expressed in the following general form[1]:

$$
\begin{aligned}
&\text{Minimize } F_i(\bar{X}) \quad i = 1, 2, \ldots, M \\
&\text{subject to } g_j(\bar{X}) \leq 0 \quad j = 1, 2, \ldots, J \\
&h_k(\bar{X}) = 0 \quad k = 1, 2, \ldots, K
\end{aligned}
\tag{1.1}
$$

---

[1] Optimization problems can be maximization problems also with the inequalities expressed as the other way around.

© The Author(s) 2023
J. C. Bansal et al., *Sine Cosine Algorithm for Optimization*,
SpringerBriefs in Computational Intelligence,
https://doi.org/10.1007/978-981-19-9722-8_1

where $F_i(\bar{X})$ is referred as objective function or cost function in Eq. (1.1) and M denotes the number of objective functions in given optimization problem. When $M = 1$, optimization problem is termed as single-objective optimization problem, and when $M > 1$, optimization problem is referred as multi-objective optimization problem. $g_j(\bar{X})$ is called inequality constraints, s.t. $1 \leq j \leq J$, and J denotes the number of inequality constraints. $h_k(\bar{X})$ is equality constraints, s.t. $1 \leq k \leq K$, and K denotes the number of equality constraints.

$F_i(\bar{X})$, $g_j(\bar{X})$, and $h_k(\bar{X})$ are functions of the vector $\bar{X} = (x_1, x_2, \ldots x_n) \in S$. $\bar{X}$ is called decision (or design) vector, and the components $x_i \in \bar{X}$ are called decision (or design) variables. S is referred as decision space (or design space), and it can be discrete, continuous, or combination of both. Based on the underlying applications, the term design or decision (vector, variable, and space) is used interchangeably. In this book, the terms decision vector, decision variables, and decision space will be used in all further discussions.

The optimization problem given by Eq. (1.1) describes a decision problem, where we are required to find the "optimal" decision vector X out of all possible vectors in the decision space (S). The process of optimizing (maximizing or minimizing) the objective function of an optimization problem by determining the optimal values of the decision variables involved in it is called optimization. There are different categories of optimization problems. This categorization of optimization problems can be done in several ways, e.g., based on the number of objective functions, the nature of the objective functions, and the nature of the constraints. For instance, if a problem involves exactly one objective function, it is called a single-objective optimization problem. If the number of objective functions is at least two or more than two, it is referred as a multi-objective optimization problem. An optimization problem can also be categorized as a real, discrete, or mixed-integer problem based on whether the underlying decision variables are real, discrete, or mixed-integer type, respectively. When there are no conditions or constraints on the decision variables, the optimization problem is called unconstrained; otherwise, it is termed as constrained optimization problem. A detailed categorization of various optimization problems is presented in Table 1.1. For more specialized discussions of the mentioned categories, an interested reader can refer to two textbooks, Optimization for Engineering Design Algorithms and Examples by [1] and Operations research: an introduction by [2].

Different optimization methods (or techniques) are available in the literature to address various types of optimization problems as mentioned in Table 1.1. However, selecting a suitable optimization method for an optimization problem is a challenging task, as there are no general guidelines for algorithm selection for a given optimization problem. Moreover, there is no efficient general algorithm for solving non-deterministic polynomial time hard or NP-hard problems. In general, optimization methods can be classified into the following two types:

1. **Traditional (deterministic) methods**: Traditional optimization methods start from a randomly chosen initial solution and use specific deterministic rules for changing the solutions' position in the search space. Most of these methods utilize the gradient information of the objective function. The initial solutions always

Table 1.1 Classification of optimization problems

Classification criterion	Optimization problem	Features
Nature of objective function and/or constraints	Linear	Linear objective function and constraints
	Nonlinear	Nonlinear objective unction and/or constraints
	Convex	Convex objective function and feasible set
	Quadratic	Quadratic objective function and linear constraints
	Stochastic	Probabilistically determined problem variables and/or parameters
	Deterministic	Decision variables and/or parameters are known accurately
	Non smooth	Either objective function or the constraints, or both, are not differentiable
Nature of the search space	Discrete	Discrete decision variables
	Continuous	Real decision variables
	Mixed integer	Both real and integer decision variables
Nature of the optimization problem	Dynamic	Objective function varying with time
	Multi-objective	More than one objective function
	Single-objective	Exactly one objective function
Existence of constraints	Constrained	At least one constraint is involved
	Unconstrained	No constraints

follow the same path for the same starting position and converge to the fixed final position, irrespective of the number of runs. These methods provide a mathematical guarantee that a given optimization problem can be solved with a required level of accuracy within a finite number of steps. There exist sufficient literature on traditional optimization methods where different methods are capable of handling various types of optimization problems. Based on the type of problem, traditional optimization techniques may be identified as methods for solving the linear programming problems (LPP), nonlinear programming problems (NLPP), and specialized programming problems. However, traditional methods sometimes fail to handle optimization problems. Usually, these methods rely on the properties like continuity, differentiability, smoothness, and convexity of the objective function and constraints (if any). The absence of any of these properties makes traditional methods incapable of handling such optimization problems. Moreover, there are optimization problems for which no information is available about the objective function; these problems are referred as the black-box optimization problem. Traditional optimization methods or deterministic methods also fail to handle such black-box problems.

Combinatorial optimization problems such as traveling salesman problem, N-vortex problem, and halting problem are non-deterministic polynomial hard (or NP-hard) problems. Traditional optimization methods are incapable of solving these NP-hard problems within a polynomial-bound time and require an exponential time. The time complexity of the traditional methods makes these methods impractical to use. The failure of the deterministic or conventional methods inspired researchers to look for some non-deterministic or unconventional methods, which are statistically reliable, fast, and robust in dealing with a larger class of optimization problems. Stochastic methods are part of these unconventional methods, which have partially proven their superiority over traditional methods in terms of robustness, computational cost-effectiveness, and speed. However, the wide applicability of stochastic methods comes at the core of reliability. Stochastic methods are discussed in detail in the next section.

2. **Stochastic (non-deterministic) methods**: Stochastic or non-deterministic optimization methods contain inherent components of randomness and are iterative in nature. These methods utilize stochastic equations which are based on different stochastic processes and utilize different probability distributions. The stochastic nature of these equations governs the path of the solutions in the search space. In different runs of these algorithms, a solution can follow different paths, despite having a fixed initial position.

Stochastic optimization methods do not always guarantee convergence to a fixed optimal position in the search space. In fact, these methods look for near optimal solution in a predefined fixed number of iterations. N number of independent runs are simulated to ensure a statistical reliability to these methods, and in general, the number of runs $N = 30$ or 51 is used to support the claim of near optimal solution. The trade-off for sacrificing the optimal solution by stochastic methods is the fast convergence speed, low computational cost, and less time complexity. Random number generators or pseudo-random number generators play an important role in the success of the stochastic methods. A brief classification of optimization techniques and their methods are illustrated in the (Fig. 1.1).

Stochastic methods are a broad area of study. These methods are based on different stochastic processes, and discussing about all these methods and techniques is beyond the scope of this book. An interested reader can refer to any advanced book like stochastic optimization by [3] for strengthening their knowledge about the subject. However, this book provides a piece of decent information about the stochastic techniques and focuses on the meta-heuristic algorithms, particularly the sine cosine algorithm (SCA) [4]. Meta-heuristic algorithms are one class of the stochastic methods. But before discussing about meta-heuristic algorithms, let us discuss first about heuristic algorithms.

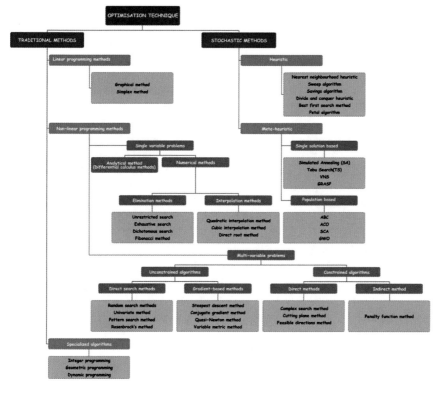

Fig. 1.1 Classification of optimization technique

The word heuristic means 'to find' or 'to discover by trial and error'. A heuristic technique or simply a heuristic is an experience-based approach that compromises accuracy, optimality, or precision for speed to solve a problem faster and more efficiently. In layman's language, a thumb rule, an intelligent guess, an intuitive judgment, or common sense can be considered as metaphors for the word heuristics. Random search algorithms, divide and conquer algorithm, nearest neighbor heuristic, savings algorithm, and best first search method are some examples of heuristic algorithms. The heuristic algorithms are known to be very specific in their search process for the solutions and are problem specific.

The word meta—means 'beyond' or 'higher level', and meta-heuristics algorithms are higher versions of heuristics algorithms. Meta-heuristic algorithms are advanced optimization algorithms and also known as modern optimization techniques. These algorithms utilize more information of the search process and less or no information of the problem; i.e., these algorithms are actually problem independent. Because of their negligible dependency on the objective function in an optimization problems, meta-heuristic algorithms are well-equipped in handling complex optimization problems and are applicable to a wider class of problems. Meta-heuristics or meta-heuristic algorithms are fast, efficient, and robust in handling highly nonlinear, non-differentiable, and even black-box optimization problems. Inexpensive computational cost or less computational complexity is one of the major advantages of using meta-heuristic algorithms.

The basic idea behind the working of meta-heuristic algorithms is simple and easy to implement. Meta-heuristic algorithms start the search process by randomly initializing a finite set of representative solutions in the search space. These solutions are also referred as particles, search agents, or individuals and will be used interchangeably throughout the text depending upon the context. The finite set containing the representative solutions is referred as the 'population'. The initial position of the search agents is evaluated using the given objective function. The population iteratively updates the positions of its search agents to look for the optimal solution in the search space. The position update mechanism of search agents during the search can be considered as the soul of the meta-heuristic algorithm.

In random search algorithms, search agents update their position randomly and do not utilize any information from each other. But, in meta-heuristic algorithms, information sharing between search agents is one of the most important components. The algorithm evaluates the position of search agents in the search space using objective function value, and a fitness value is assigned. The fitness of a search agent is the value of the objective function at its position. Search agents lying near to optimum location have better fitnesses, and agents far from the optimum have poor fitness values. Better search agents communicate about the their position to other agents, and other agents try to follow the direction of better agents to improve their fitness values.

Mathematically speaking, suppose $S \subseteq \mathbb{R}^D$ is a D-dimensional subspace of the $\mathbb{R}^D$, and the population size is N. If $X_i = (x_1, x_2, \ldots, x_D) \in S$ is the current position vector of the ith $(1 \leq i \leq N)$ search agent in the search space S, then a simple position update mechanism can be described by Eq. (1.2);

$$X_i^{\text{new}} = X_i^{\text{curr.}} + \overline{h}_i \tag{1.2}$$

where $\overline{h}_i$ is a D-dimensional step vector determining the magnitude of the step length and direction of the position update for ith agent. The addition (+) in Eq. (1.2) is vector addition or component-wise addition. Step vector $\overline{h}$ is produced by meta-heuristic algorithms may contain components of best search agent's position, worst search agent's position, the mean of the positions, and some random scaling factors. For instance, particle swarm optimizer (PSO) utilizes a position update mechanism similar to that mentioned in Eq. (1.2).

One more major position update mechanism can be realized by changing the components of the position vector in the search space. Suppose $X_i = (x_{i,1}, x_{i,2}, \ldots, x_{i,D})$ is the position vector of the ith search agent in the search space S. If we replace some components $x_{i,j}$, where $1 \leq j \leq D$, by different values, say $u_{i,j}$, such that $u_{i,j} \neq x_{i,j}$, the position of X_i will be changed. Similarly, if a nontrivial permutation operator is applied on the components of X_i, the position of X_i can be updated. Genetic algorithms are one class of algorithms utilizing similar technique to update the position of search agents in the search space. A hybrid of these two position update mechanisms can also be employed by some available meta-heuristic algorithms. For example, differential evolution (DE) exploits the combination of both these techniques. The latest development in the field of meta-heuristic algorithms is utilizing more advanced versions of these position update mechanisms, although the underlying idea is the same as discussed above.

In meta-heuristic algorithms, position update mechanisms are dynamic in nature and utilize the information from the ongoing optimization process. In these algorithms, using any large step sizes (or large changes in the position) of the search agents can hamper the convergence of the algorithm, and very small step sizes (or very small changes in the position) of the search agents lead to stagnation and slow speed. Stagnation is the phase of any meta-heuristic algorithm, when search agents in the search space lose their diversity and converge to a local optimal solution. Both of these extreme situations are not good for any optimization algorithm. So, in any meta-heuristic algorithm achieving a fine balance between the large steps and small steps is of paramount importance. This process of achieving a fine balance between step sizes is referred as 'exploitation versus exploration' or 'intensification versus diversification'.

In the exploitation phase, the algorithm utilizes very small step sizes to extensively cover the local region of the search space where the optimum can lie. Search agents make very small changes in their position to scan the local region of the search space thoroughly. However, its disadvantage is that it makes the convergence speed slow. On the other hand, exploration refers to the capability of the algorithm to cover the large size of search space efficiently and maintain the diversity in the population of the search agents. Therefore, exploration can be considered as a searching process on a global scale. Large step sizes make exploration less prone to stuck in the local optimum locations and help in finding the region of the global optimum. The major disadvantage of high exploration rate is that it can skip the global optima and converge prematurely. So, the optimal balance between exploration and exploitation is a very critical component of the algorithm.

The advancements in the literature of meta-heuristic algorithms have grown significantly in the recent past. There are various classifications available. For instance, meta-heuristic algorithms can be categorized based on their source of inspiration, their country of origin, whether they originate from natural or some artificial phenomenon, and whether they start with multiple solutions or single solutions [5]. For a good overview of the classification of meta-heuristic algorithms, an interested reader can refer [6–8]. Based on the number of representative solutions in the search space, i.e., multiple solutions and single solutions, meta-heuristic algorithms can be classified into two categories: population-based and single solution-based. The population-based meta-heuristic algorithms begin with a set of random representative solutions, which are then improved iteratively until the termination criterion is satisfied. Some of the popular meta-heuristic algorithms are particle swarm optimization (PSO) [9], artificial bee colony (ABC) [10], sine cosine algorithm (SCA) [4], ant colony optimization (ACO) [11], differential evolution (DE) [12], genetic algorithms (GA) [13], gravitational search algorithm (GSA) [14], teaching–learning-based optimization (TLBO) [15], gray wolf optimization algorithm (GWO) [16], spider monkey optimization (SMO) [17], and many others. Single-solution-based algorithms generate a single solution and improve the solution until the termination condition is satisfied. Methods like simulated annealing (SA) [18], noising method (NM) [19], the tabu search (TS) [20], variable neighborhood search (VNS) [21], and the greedy randomized adaptive search procedure (GRASP) [22] method fall under this category. Population-based meta-heuristic algorithms are preferred over single-solution-based algorithms because of their robust exploration capabilities, i.e., checking multiple points in the search space simultaneously saving time and resources and improving the probability of reaching the global optima.

Population-based meta-heuristic algorithms can be studied under two major categories of evolutionary algorithms (EAs) and swarm intelligence (SI)-based algorithms. The underlying principles and working of these algorithms are similar but their source of inspiration is different. Brief detail about these algorithms is mentioned below:

1. **Evolutionary Algorithms**: Evolutionary algorithms (EAs) are inspired by the natural evolutionary process. The structure of the evolutionary algorithm is based on the Darwinian theory related to the biological evolution of species and the survival of the fittest principle. In EAs, search agents or solutions evolve iteratively using three major operators—selection, mutation, and crossover (or recombination). The family of evolutionary algorithms comprises genetic algorithms (GA), evolution strategies, differential evolution (DE), genetic programming (GP), biogeography-based optimization [23], evolutionary programming, etc.

2. **Swarm Intelligence (SI)-Based Algorithms**: Beni and Wang [24] coined the phrase "swarm intelligence" (SI) in 1993 to describe the cooperative behavior of robotic systems. SI is an important branch of artificial intelligence in which complex, autonomous, and decentralized systems are studied. Swarm can be described as a collection of simple entities which corporate with each other to execute complex tasks, for example the collective behaviors of social ants, cooperation of honey bees, etc. Swarm of simple autonomous agents interact with each other and demonstrate intelligent traits such as the ability to make decisions and adaptability to change when aggregated together. Meta-heuristic algorithms in which the autonomous agents work together to find the optimal solution and do not involve evolutionary operators are termed as swarm intelligence (SI)-based algorithms. Some well-known algorithms under this category are PSO, ABC, ACO, GSA, SCA, and TSA.

In the mid-90s, EA and SI algorithms were studied under the single category of evolutionary computing, because of their similarities, such as using a population of the solution and their stochastic nature. Although the underlying motivation of these algorithms is different. In evolutionary algorithms, new solutions emerge and old solutions die in the optimization process, while in SI algorithms, old solutions are improved iteratively, and no old solution die in the optimization process. Researchers noticed this difference, and consequently, more academic research on swarm intelligence was published in the international academic journals, making the field of SI-based algorithms more popular and applied.

Meta-heuristic algorithms can also be further categorized based on their source of inspiration from different fields of sciences like life science, physics, mathematics, etc. Some of the major categories of the algorithms falling in these categories are discussed below:

1. **Life Science-Based Algorithms**: Life science concerns with the study of living organisms, from single cells to human beings, plants, microorganisms, and animals. Meta-heuristic algorithms that take inspiration from the species of birds, animals, fishes, bacteria, microorganisms and viruses, plants, trees, fungi, and human organs, like kidney, heart, or disease treatment methods, such as chemotherapy, come under this category. This category can be further classified as fauna-based, flora-based, and organ-based [25]. A few examples of meta-heuristic algorithms that fall under this category are GWO, PSO, ABC, ACO, artificial plant optimization algorithm [26], root tree optimization algorithm [27], chemotherapy science algorithm [28], kidney-inspired algorithm [29], and heart algorithm [30].

2. **Physical Science-Based Algorithms**: Physical science includes physics, chemistry, astronomy, and earth science. Algorithms that imitate the behavior of physical or chemical phenomena, such as electromagnetism, water movement, electric charges/ions, chemical reactions, gaseous particle movement, celestial bodies, and gravitational forces are grouped under this category. Some popular physical science-based algorithms are black hole optimization [31], crystal energy optimization algorithm [32], ions motion optimization algorithm [33], galaxy-based search algorithm [34], gravitational search algorithm, simulated annealing, and atmosphere clouds model [35].

3. **Social Science-Based Algorithms**: Social science deals with the behavior of humans and the functioning of human colonies. It covers exciting fields like human geography, psychology, economics, political science, history, and sociology. Meta-heuristic algorithms under this category have drawn inspiration from humans' social and individual conduct. The principles of leadership, decision-making, economics, and political or competitive ideologies are some of the concepts that have served as the sources of inspiration. Some have even borrowed metaphors from how humans rule territories and economic systems. Some of the algorithms that fall under this category are ideology algorithm [36], greedy politics optimization algorithm [37], parliamentary optimization algorithm [38], imperialist competitive algorithm [39], social emotional optimization algorithm [40], anarchic society optimization [41], brain storm optimization algorithm [42], and teaching–learning-based optimization (TLBO) [15]. This category also includes algorithms that are inspired by the activities or events introduced by humans, such as the soccer league competition algorithm [43], league championship algorithm [44], and tug of war optimization [45].

4. **Mathematics-Based Algorithms**: This category includes algorithms inspired by mathematical models and mathematical equations. Some of the examples of the mathematics-based algorithms are gradient-based optimizer (GBO) [46], Runge–

Kutta optimization (RUN) [47], tangent search algorithm (TSA) [48], sine cosine algorithm (SCA), differential evolution (DE), and stochastic fractal search (SFS) [49].

Population-based meta-heuristic methods are gaining increasing attention from researchers in the scientific community over the recent past. These methods are more efficient and cost-effective in solving complex problems. The major advantages of population-based meta-heuristic algorithms are summarized here:

(1) Population-based meta-heuristics are easy to implement and enable better exploration of the search space than single-solution-based algorithms.
(2) They initiate the search process with multiple randomly generated solutions in the search space. The presence of multiple solutions in the search space enables solutions to share information about the search space with other solutions and prevents premature convergence in a local optimal region.
(3) Since meta-heuristic frameworks follow general principles, which makes population-based meta-heuristic algorithms easily applicable on a wide variety of real-life optimization problems.
(4) In general, meta-heuristics do not rely on the information about the optimization problem formulation (such as the requirement of constraints or objective functions to be linear, continuous, differentiable, convex, etc.), they are more robust and optimization-friendly.

Modern optimization techniques like particle swarm optimization (PSO) [9], artificial bee colony (ABC) [10], differential evolution (DE) [12], firefly algorithm (FA) [50], ant colony optimization (ACO) [11], black hole optimization (BHO) [31], teaching–learning-based optimization (TLBO) [15], genetic algorithm (GA) [13], spider monkey optimization (SMO) [17], gravitational search algorithm (GSA) [14], gray wolf optimization algorithm (GWO) [16], sine cosine algorithm (SCA) [4], have emerged as popular methods for tackling challenging problems in both industries and academic research. Sine cosine algorithm (SCA) is a new mathematical concept-based meta-heuristic algorithm. SCA uses trigonometric functions (sine and cosine) to update the position of the search agents in the search space. It has shown promising results in solving various optimization problems. SCA was introduced by Mirjalili [4] to develop a user-friendly, robust, effective, efficient, and easy-to-implement algorithm that demonstrates decent capabilities in exploring and exploiting the search space. This book is dedicated to the study of sine cosine algorithm (SCA) and its applications. The motive of this book is to discuss and present a fair amount of information about the sine cosine algorithm, which might be helpful for fellow readers who wish to work in the field of meta-heuristic algorithms. The basic SCA algorithm, its variants, and its applications are discussed in the subsequent chapters of the book.

Practice Exercises

1. Discuss the difference between traditional optimization algorithms and meta-heuristic algorithms.
2. Describe the shortcomings of traditional optimization techniques.
3. Write a short note on challenges in the meta-heuristic algorithms.
4. Discuss the difference between evolutionary algorithms and swarm intelligence algorithms.

References

1. K. Deb, *Optimization for Engineering Design: Algorithms and Examples* (PHI Learning Pvt. Ltd., 2012)
2. H.A. Taha, *Operations Research: An Introduction*, vol. 790 (Pearson/Prentice Hall, Upper Saddle River, NJ, 2011)
3. S.K.J. Schneider, *Stochastic Optimization* (Springer, 2006)
4. S. Mirjalili, SCA: a sine cosine algorithm for solving optimization problems. Knowl.-Based Syst. **96**, 120–133 (2016)
5. H.R. Moshtaghi, A.T. Eshlaghy, M.R. Motadel, A comprehensive review on meta-heuristic algorithms and their classification with novel approach. J. Appl. Res. Ind. Eng. **8**(1), 63–89 (2021)
6. A.E. Ezugwu et al., Metaheuristics: a comprehensive overview and classification along with bibliometric analysis. Artif. Intell. Rev. **54**(6), 4237–4316 (2021)
7. D. Molina et al., Comprehensive taxonomies of nature- and bio-inspired optimization: inspiration versus algorithmic behavior, critical analysis recommendations. Cogn. Comput. **12**(5), 897–939 (2020)
8. H. Stegherr, M. Heider, J. Hähner, Classifying metaheuristics: towards a unified multi-level classification system. Nat. Comput. 1–17 (2020)
9. J. Kennedy, R. Eberhart, Particle swarm optimization, in *Proceedings of ICNN'95-International Conference on Neural Networks*, vol. 4 (IEEE, 1995), pp. 1942–1948
10. D. Karaboga, *An idea based on honey bee swarm for numerical optimization* (Technical report-tr06). Erciyes University, Engineering Faculty, Computer, 2005
11. M.D.L.M. Gambardella, M.B.A. Martinoli, R.P.T. Stützle, Ant colony optimization and swarm intelligence, in *5th International Workshop* (Springer, 2006)
12. R. Storn, K. Price, Differential evolution—a simple and efficient heuristic for global optimization over continuous spaces. J. Glob. Optim. **11**(4), 341–359 (1997)
13. M. Mitchell, *An Introduction to Genetic Algorithms* (MIT Press, 1998)
14. E. Rashedi, H. Nezamabadi-Pour, S. Saryazdi, GSA: a gravitational search algorithm. Inf. Sci. **179**(13), 2232–2248 (2009)
15. R. Venkata Rao, V.J. Savsani, D.P. Vakharia, Teaching-learning-based optimization: an optimization method for continuous non-linear large scale problems. Inf. Sci. **183**(1), 1–15 (2012)
16. S. Mirjalili, S.M. Mirjalili, A. Lewis, Grey wolf optimizer. Adv. Eng. Softw. **69**, 46–61 (2014)
17. J.C. Bansal et al., Spider monkey optimization algorithm for numerical optimization. Memet. Comput. **6**(1), 31–47 (2014)
18. S. Kirkpatrick, C. Daniel Gelatt, Jr., M.P. Vecchi, Optimization by simulated annealing. Science **220**(4598), 671–680 (1983)
19. I. Charon, O. Hudry, The noising method: a new method for combinatorial optimization. Oper. Res. Lett. **14**(3), 133–137 (1993)

20. F. Glover, Future paths for integer programming and links to artificial intelligence. Comput. Oper. Res. **13**(5), 533–549 (1986)
21. N. Mladenovic, A variable neighborhood algorithm—a new metaheuristic for combinatorial optimization. Papers presented at Optimization Days, vol. 112 (1995)
22. T.A. Feo, M.G.C. Resende, Greedy randomized adaptive search procedures. J. Glob. Optim. **6**(2), 109–133 (1995)
23. D. Simon, Biogeography-based optimization. IEEE Trans. Evol. Comput. **12**(6), 702–713 (2008)
24. G. Beni, J. Wang, Swarm intelligence in cellular robotic systems, in *Robots and Biological Systems: Towards a New Bionics?* (Springer, 1993), pp. 703–712
25. A. Tzanetos, G. Dounias, A comprehensive survey on the applications of swarm intelligence and bio-inspired evolutionary strategies, in *Machine Learning Paradigms* (2020), pp. 337–378
26. Z. Zhao, Artificial plant optimization algorithm for constrained optimization problems, in *2011 Second International Conference on Innovations in Bio-Inspired Computing and Applications* (IEEE, 2011), pp. 120–123
27. Y. Labbi et al., A new rooted tree optimization algorithm for economic dispatch with valve-point effect. Int. J. Electr. Power Energy Syst. **79**, 298–311 (2016)
28. M.H. Salmani, K. Eshghi, A metaheuristic algorithm based on chemotherapy science: CSA. J. Optim. **2017** (2017)
29. N.S. Jaddi, J. Alvankarian, S. Abdullah, Kidney-inspired algorithm for optimization problems. Commun. Nonlinear Sci. Numer. Simul. **42**, 358–369 (2017)
30. A. Hatamlou, Heart: a novel optimization algorithm for cluster analysis. Prog. Artif. Intell. **2**(2), 167–173 (2014)
31. A. Hatamlou, Black hole: a new heuristic optimization approach for data clustering. Inf. Sci. **222**, 175–184 (2013)
32. X. Feng, M. Ma, H. Yu, Crystal energy optimization algorithm. Comput. Intell. **32**(2), 284–322 (2016)
33. B. Javidy, A. Hatamlou, S. Mirjalili, Ions motion algorithm for solving optimization problems. Appl. Soft Comput. **32**, 72–79 (2015)
34. H. Shah-Hosseini, Principal components analysis by the galaxy-based search algorithm: a novel metaheuristic for continuous optimisation. Int. J. Comput. Sci. Eng. **6**(1–2), 132–140 (2011)
35. G.-W. Yan, Z. Hao, J. Xie, A novel atmosphere clouds model optimization algorithm. J. Comput. (Taiwan) **24**(3), 26–39 (2013)
36. T.T. Huan et al., Ideology algorithm: a socio-inspired optimization methodology. Neural Comput. Appl. **28**(1), 845–876 (2017)
37. J.S.M. Lenord Melvix, Greedy politics optimization: metaheuristic inspired by political strategies adopted during state assembly elections, in *2014 IEEE International Advance Computing Conference (IACC)* (IEEE, 2014), pp. 1157–1162
38. A. Borji, M. Hamidi, A new approach to global optimization motivated by parliamentary political competitions. Int. J. Innov. Comput. Inf. Control **5**(6), 1643–1653 (2009)
39. E. Atashpaz-Gargari, C. Lucas, Imperialist competitive algorithm: an algorithm for optimization inspired by imperialistic competition, in *2007 IEEE Congress on Evolutionary Computation* (IEEE, 2007), pp. 4661–4667
40. Y. Xu, Z. Cui, J. Zeng, Social emotional optimization algorithm for nonlinear constrained optimization problems, in *International Conference on Swarm, Evolutionary, and Memetic Computing* (Springer, 2010), pp. 583–590
41. A. Ahmadi-Javid, Anarchic society optimization: a human-inspired method, in *2011 IEEE Congress of Evolutionary Computation (CEC)* (IEEE, 2011), pp. 2586–2592
42. Y. Shi, Brain storm optimization algorithm, in *International Conference in Swarm Intelligence* (Springer, 2011), pp. 303–309
43. N. Moosavian, B.K. Roodsari, Soccer league competition algorithm: a novel metaheuristic algorithm for optimal design of water distribution networks. Swarm Evol. Comput. **17**, 14–24 (2014)

44. A.H. Kashan, League championship algorithm: a new algorithm for numerical function optimization, in *2009 International Conference of Soft Computing and Pattern Recognition* (IEEE, 2009), pp. 43–48
45. A. Kaveh, A. Zolghadr, A novel meta-heuristic algorithm: tug of war optimization. Iran Univ. Sci. Technol. **6**(4), 469–492 (2016)
46. I. Ahmadianfar, O. Bozorg-Haddad, X. Chu, Gradient-based optimizer: a new metaheuristic optimization algorithm. Inf. Sci. **540**, 131–159 (2020)
47. I. Ahmadianfar et al., RUN beyond the metaphor: an efficient optimization algorithm based on Runge Kutta method. Expert Syst. Appl. **181**, 115079 (2021)
48. A. Layeb, Tangent search algorithm for solving optimization problems. Neural Comput. Appl. **34**(11), 8853–8884 (2022)
49. H. Salimi, Stochastic fractal search: a powerful metaheuristic algorithm. Knowl.-Based Syst. **75**, 1–18 (2015)
50. X.-S. Yang, Firefly algorithms for multimodal optimization, in *International Symposium on Stochastic Algorithms* (Springer, 2009), pp. 169–178

Chapter 2
Sine Cosine Algorithm

Sine cosine algorithm (SCA) [1] is relatively a new algorithm, in the field of meta-heuristic algorithms. SCA is a population-based probabilistic search method that updates the position of search agents in the population using simple concept of trigonometric functions sine and cosine. SCA algorithm is inspired from the periodic property of the sine and cosine functions. The periodicity of the sine and cosine function in the range $[-1, 1]$ provides great capacity to exploit the search space and helps in maintaining a fine balance between exploration and exploitation. In previous Chap. 1, we have already discussed about the criticality of the exploration and exploitation capabilities of any meta-heuristic algorithm.

Trigonometric functions sine and cosine are periodic functions with a period of 2π. The range of both the functions is $[-1, 1]$. The variation of these functions between -1 and $+1$ offers a great capacity to scan the local regions in the search space containing global optima and provides the required diversity to the search agents in the search space. Like any other meta-heuristic algorithm, SCA is a random search technique that is not a problem-dependent technique, and it does not require gradient information of the objective function. SCA is a population-based probabilistic search technique, it starts the search process with multiple randomly initialized representative solutions or search agents in the search space, and updates the position of search agents toward or away from the best candidate solution using a mathematical model based on the sine and cosine functions.

Sine cosine algorithm (SCA) is becoming increasingly popular over the past few years. The SCA's popularity is evident from the SCA-related papers published in several reputed journals over the time. Figure 2.1 gives a fair idea about the number of research publications in the last six years. All these research publications contain the sine cosine algorithm in their title, abstract, and keywords. The upward trend of increasing interest in the SCA is due to its robust optimization capabilities and simplicity in implementation. It has successfully been applied to tackle the complex real-world optimization problems of different scientific disciplines, such as electrical engineering, control engineering, combinatorial problems, machine learning,

© The Author(s) 2023
J. C. Bansal et al., *Sine Cosine Algorithm for Optimization*,
SpringerBriefs in Computational Intelligence,
https://doi.org/10.1007/978-981-19-9722-8_2

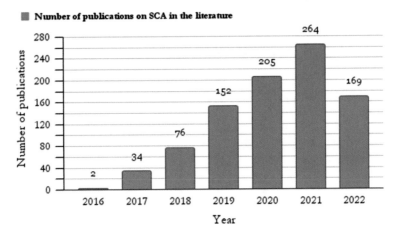

Fig. 2.1 Number of papers published on sine cosine algorithm in the title, abstract and keywords. *Source* SCOPUS database, till July 2022

robotics, supply chain problems, and environmental science problems, to name a few. The spectrum of SCA applications is broad and spans over diverse fields of science and technology.

The purpose of this chapter is to serve the readers about the insights of the basic sine cosine algorithm. The present chapter covers the fundamentals of the sine cosine algorithm with a step-by-step implementation of the algorithm. A simple numerical example with a MATLAB code is added for the readers to fully understand the procedure involved in the working of the sine cosine algorithm. The strengths and weaknesses of the SCA algorithm are also discussed in this chapter to give readers a fair idea on the utility of the algorithm in the different fields of scientific research. The present chapter will encourage the researchers to modify the original SCA and implement it to solve various optimization problems.

The chapter is organized as follows: Sect. 2.1 describes the basic principles of the SCA algorithm and its pseudo-code. The control parameters involved in the SCA algorithm and the impact of these control parameters on the performance of the algorithm are discussed in Sect. 2.2. A simple numerical example explaining the computational procedure of the basic SCA algorithm is described in Sect. 2.3. The MATLAB code of the SCA algorithm handling the numerical example mentioned in Sect. 2.4, and for summarizing the chapter, concluding remarks are given in Sect. 2.5.

2.1 Description of the Sine Cosine Algorithm (SCA)

Similar to any other population-based optimizers, sine cosine optimization process begins with randomly initializing a set of representative solutions or search agents in the search space. The set containing all search agents is also referred as the population.

In the population, each search agent can be treated as a vector in a d-dimensional search space. Search agents in the search space update their position with the help of stochastic equations containing the trigonometric sine and cosine functions.

The population in the search space is randomly initialized within the search space bounds using Eq. (2.1). The ith search agent $X_i = (X_{i1}, X_{i2} \ldots X_{id})$ is initialized using the following equation:

$$X_{ij} = X_{ij}^{lb} + \text{rand}() * \left(X_{ij}^{ub} - X_{ij}^{lb} \right), \quad j = 1 : d, \ i = 1 : \text{Np} \qquad (2.1)$$

where X_{ij} represents the jth dimension of the ith solution, X_{ij}^{lb} and X_{ij}^{ub} denote the lower bound and upper bound of the ith solution in the jth dimension of the search space, respectively. The function rand() generates uniformly distributed random numbers in the range [0, 1], and Np denotes the number of the search agents in the population, i.e., the population size.

The next step after initializing the population in the search space is to update the position of each search agent to look for the optimal solution. For this purpose, the position of the each agent is evaluated using the underlying objective function, and based on the optimization criteria, a fitness value or a goodness value is assigned to each agent. The search agent with the highest fitness is considered as the best search agent, and the position of the best search agent is referred as the destination point. After locating the destination point, other search agents update their position in the search space (or design space) using the destination point as a reference. The following equations are position update equations:

$$X_{ij}^{t+1} = X_{ij}^t + r_1 \times \sin(r_2) \times \left| r_3 \times P_g^t - X_{ij}^t \right| \qquad (2.2)$$

$$X_{ij}^{t+1} = X_{ij}^t + r_1 \times \cos(r_2) \times \left| r_3 \times P_g^t - X_{ij}^t \right| \qquad (2.3)$$

where, $j = 1 : d$, and $i = 1 : \text{Np}$.

$X_i^t = (X_{i1}^t, X_{i2}^t \ldots X_{id}^t)$ denotes the position of the ith search agent in the tth iteration. $P_g^t = (P_{g1}^t, P_{g2}^t, \ldots P_{gd}^t)$ is the gth search agent having the best fitness and considered as the destination point at tth iteration. $|.|$ represents the modulus operator. r_1 is a function of iteration counter t, calculated using Eq. (2.4), here b is a constant parameter and T denotes the maximum number of iterations. r_2 and r_3 are uniformly distributed random numbers generated using Eqs. (2.5) and (2.6), respectively:

$$r_1 = b - b \times \left(\frac{t}{T} \right) \qquad (2.4)$$

$$r_2 = 2 \times \pi \times \text{rand}() \qquad (2.5)$$

$$r_3 = 2 \times \text{rand}() \qquad (2.6)$$

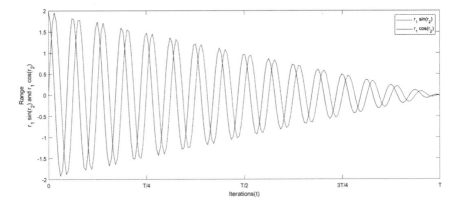

Fig. 2.2 Trajectory of $r_1 \sin(r_2)$ and $r_1 \cos(r_2)$ considering $c = 2$

A proper balance between exploration and exploitation is of paramount importance in any population-based optimization algorithm. At the initial phase of the optimization process or in early iterations, an algorithm should focus on the exploration process to sufficiently scan the design space. At a later stage or in later iterations, the algorithm should use the exploitation process to search the promising local regions to find the global optimal location in the search space and guarantee convergence. So with the increasing number of iterations, the exploration ability of the algorithm should decrease, while exploitation capabilities should increase. In sine cosine algorithm, the control parameter r_1 is responsible for maintaining the balance between the exploration and exploitation process. This parameter ensures a smooth transition from the exploration phase to exploitation phase during the search. The control parameter r_1 is linearly decreasing function of iteration counter t, which a linearly reduces the value of the constant parameter b. The trigonometric functions sine and cosine in Eqs. (2.2) and (2.3) are multiplied by the control parameter r_1. That means, the range of these terms is dependent on the value of the control parameter r_1. The value of r_1 is dependent on the constant parameter b, whose value is linearly decreasing with the increasing number of iterations. So, by controlling the value of the constant parameter b, SCA algorithm controls the range of the terms $r_1 \cdot \sin(r_2)$ and $r_1 \cdot \cos(r_2)$. The trajectory of the range of $r_1 \cdot \sin(r_2)$ and $r_1 \cdot \cos(r_2)$ during the search process is illustrated in Fig. 2.2.

Moreover, it is not a difficult observation to make for a reader that the control parameter r_1 works as a scaling factor for the step size in the position update equations given by Eqs. (2.2) and (2.3). In early iterations, larger values of r_1 is used by the SCA algorithm to perform larger movements by the search agents to explore the search space, and at later iterations, value of r_1 will decrease to perform small movements by the search agents to ensure exploitation in the potential local regions of the search space. So, the control parameter is a critical component in the SCA algorithm for maintaining a fine-tune balance between exploration and exploitation.

A detailed discussion about the control parameters associated with the SCA algorithm is presented in subsequent Sect. 2.2.

To increase the robustness of the sine cosine algorithm, two position update equations, or in other words two separate mechanisms, are used in the SCA algorithm. To determine whether Eqs. (2.2) or (2.3) should be used to update the position of the search agents, a switch probability $p(p = 0.5)$ is used, depending on a generated random number $r_4 \in [0, 1]$. If $r_4 < p$, Eq. (2.2) is used to update the position of the search agents, otherwise Eq. (2.3) is used. The following equation summarizes the above mechanism,

$$
X_{ij}^{t+1} = \begin{cases} X_{ij}^t + r_1 \times \sin(r_2) \times \left| r_3 \times P_{gj}^t - X_{ij}^t \right| & \text{if } r_4 < p \\[3mm] X_{ij}^t + r_1 \times \cos(r_2) \times \left| r_3 \times P_{gj}^t - X_{ij}^t \right| & \text{if } r_4 \geq p \end{cases}
\tag{2.7}
$$

It is evident from Eq. (2.7) that it gives 50% chance to each update equation.

The search agents in the SCA algorithm follow a nonlinear search route because of the presence of the absolute value term and the trigonometric functions sine cosine in the position update equations. Figure 2.3 illustrates the movement of the search agents with respect to the destination point (P_g) in a two-dimensional search space. It demonstrates that the search agents follow a circular path, with the best solution or destination point in the center and all other search agents positioned around it. The value of constant parameter b is taken to be 2 in the SCA algorithm that means the sine and cosine functions will operate in the range $[-2, 2]$. Each search agent updates its position either in the direction opposite to the destination point or toward anywhere between its current position and the destination point. The potential local regions where search agent X_i can move are described by dividing the circular search domain into sub-regions as shown in Fig. 2.3. The value of r_1 controls the movement of X_i, if $r_1 < 1$, then X_i moves toward destination point P_g (exploitation step), and when $r_1 \geq 1$, the search agent moves far away from the destination point P_g (exploration step).

The pseudo-code for the basic sine cosine algorithm is given in Algorithm 1, and the flowchart is shown in Fig. 2.4 to provide a concise description of the underlying working procedure of the SCA algorithm.

2.2 Parameters Associated with the SCA

The convergence speed and optimization capabilities of a population-based metaheuristic algorithm are greatly influenced by the associated parameters. The choice of the parameters' values determines the convergence rate of an algorithm. The control parameters associated with the sine cosine algorithm are r_1, r_2, r_3, and r_4.

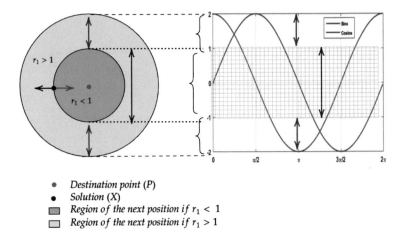

- Destination point (P)
- Solution (X)
- ▨ Region of the next position if $r_1 < 1$
- ▢ Region of the next position if $r_1 > 1$

Fig. 2.3 Impact of the parameter r_1 on the sine and cosine function or decreasing pattern

---

Algorithm 1 Sine cosine algorithm (SCA)

Initialize the population $\{X_1, X_2, \ldots, X_N\}$ randomly in the search space
Initialize the parameters associated with SCA
Calculate the objective function value for each search agent in the population
Identify the best solution obtained so far as the destination point P_g
initialize $t = 0$, where t is iteration counter
while Termination criteria is met **do**
    Calculate r_1, using Eq. (2.4) and generate the parameters r_2, r_3, r_4 randomly
    for each search agent **do**
      Update the position of search agents using Eq. (2.7)
    end for
    Update the current best solution (or destination point) P_g
    $t = t + 1$
end while
Return the best solution P_g

In SCA algorithm, the control parameter r_1 regulates both the global and local search operations. It determines whether to advance the search agents in the direction of the best solution (destination point ($r_1 > 1$)) or move the search agents away from the destination point ($r_1 < 1$) in the search space. With the increasing number of iterations, its value declines linearly from the initial parameter value 'b' to 0. This adaptive behavior of r_1 assists the SCA algorithm in ensuring the exploration behavior in early iterations and controls the exploitation behavior at later iterations.

The control parameter r_2 determines how far the search agents should travel toward or away from the destination point. Its value lies in the range $[0, 2\pi]$. The parameter r_3 in the SCA algorithm is the random weight associated with the destination point. It controls how much the destination point will contribute in updating the position of other search agents in subsequent iterations. It is a random scaling factor, and responsible for boosting ($r_3 > 1$) or lowering ($r_3 < 1$) the influence of the best

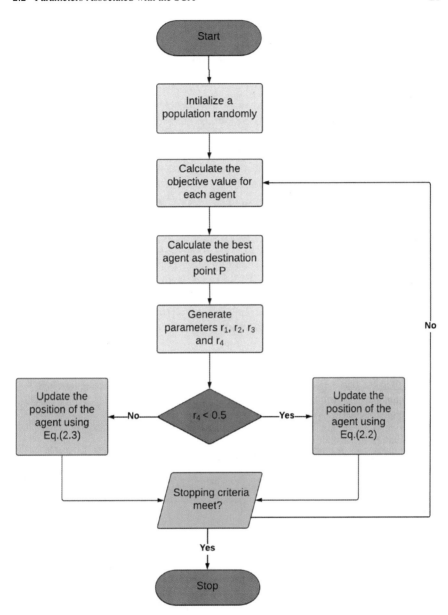

Fig. 2.4 Flowchart of SCA

solution by controlling the length of the movement. In other words, a weight greater than one indicates that the influence of the destination point is higher in finding the next position of the other search agents. On the other hand, a weight less than one indicates the lower influence of the destination point in updating the position of rest of the search agents.

The parameter r_4 is employed to randomly switch between the sine and cosine components of the position update equations. If r_4 is less than 0.5, the position update equation with the sine function is selected, and if the value of r_4 is greater than or equal to 0.5, the position update equation containing the cosine function is used. It aids the SCA's ability to avoid local optimal points in the search space and enhances the robustness of the algorithm. The value of parameter r_4 is generated using uniformly distributed random number in the range [0, 1]. Another additional important parameter in the SCA algorithm is the constant parameter 'b'. It is a preset parameter that ensures that the algorithm transit smoothly from the exploration phase to the exploitation phase. The value of the constant parameter 'b' is suggested to be 2 in the basic SCA algorithm. Like any other population-based algorithm, the performance of the SCA algorithm is also sensitive to the population size. The size of the population is a user-controlled parameter whose value is often selected on the basis of the complexity of the underlying optimization problem.

Broadly, the advantages and disadvantages of SCA can be summarized as below:

Advantages	Disadvantages
Sine cosine algorithm is a simple population-based algorithm. It is easy to implement and user-friendly	As compared to other types of problems, its performance is good for continuous optimization problems only
It has a tendency toward the best regions of the search space as it updates its position around the best solution	It lacks internal memory (i.e., it does not keep the track of previously obtained potential or best solutions)
It has a higher explorative ability as it uses four random parameters r_1, r_2, r_3 and r_4	It has a weak exploitative ability as it does not preserve the previously obtained potential solutions capable of converging to the global optimal solution
Attributing to its simple code, its speed is fast	Being a stochastic technique, it does not guarantee the global optimal solution
SCA transits smoothly from the exploration to the exploitation phase	It shows slow convergence in some of the complex optimization problems

Population-based optimizers are in great demand in the field of academic research and industrial applications. A simple and user-friendly practical optimization technique can be considered as a good optimizer if it follows certain characteristics like:

1. Optimizer should be robust, problem independent, and capable of handling black-box optimization problems.
2. The ability to locate the global optimal or near global optimal solution regardless of the complexity of the search space and modality of the objective function with a high convergence rate.

3. It should have less number of control parameters and tuning of the parameters should not be a challenging task.

Sine cosine algorithm significantly fulfills all the criteria for considering as a good optimizer. SCA has shown its robust performance capabilities in many complex, real-world optimization problems where traditional methods fail or have limited applicability. The simplicity of the SCA algorithm makes it user-friendly and simple to implement in any computer language. The less number of control parameters and adaptive nature in managing the balance between exploration and exploitation is one of the major characteristics of the SCA algorithm. The performance of SCA algorithm is exceptional in dealing with various benchmark problems. Ease of implementation, wide range of applicability, and high level of reliability make sine cosine algorithm a worthy candidate in the class of meta-heuristics.

2.3 Biases of Sine Cosine Algorithm

The major drawback of any meta-heuristic algorithm is(are) its intrinsic bias(es), or in other words, the tendency of the algorithm to accumulate solutions in a particular region(s) of the search space. For example, if the intrinsic bias of an algorithm is central bias, the algorithm will accumulate solutions in the central region, irrespective of the underlying objective function. This implies that if the objective function's true optima lies in the central region of the search space, the chances for finding the near optimal solution are favorable to the algorithm. However, on the other hand, if true optima is lying in some different regions of the search space, the chances for finding the near optimal solution will be very less; that is, the algorithm's performance will deplete on the set of objective functions in which true optima do not lie in the central region. Similarly, an algorithm may have edge bias, in which solutions accumulate in the edges of constrained search space, or axial bias, where algorithm favors distribution of solutions along any axes of the bounded search space, or any other type of biases, like exploitation bias, in which solutions accumulate around a position with no specific characteristics, demonstrating that the algorithm is over exploiting that particular region. So, the information about the intrinsic bias(es) might help the researchers better understand these stochastic optimizers' limitations.

In theory, the intrinsic characteristics of any meta-heuristic algorithm can be accessed with the help of the mathematical analysis of the algorithm. However, in practice, it is a difficult task to detect these biases of the algorithm by simply inspecting the formula. An experimental approach is suggested, in which these stochastic algorithms are assigned to optimize an impossible 'flat' problem, that is, a constant function. The problem of optimizing a constant function with the help of a stochastic algorithm can be considered impossible to solve because all the solutions in the search space are equivalent. That means the solutions of an unbiased meta-heuristic algorithm should attain positions statistically similar to a purely random search. For this, the successive positions of solutions are examined to highlight the intrinsic bias(es)

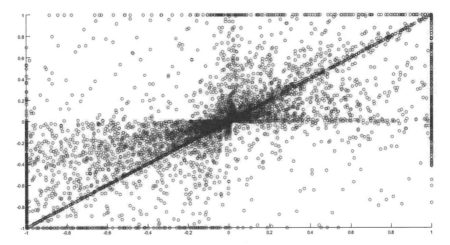

Fig. 2.5 Experiment 1—signature of SCA with 10,000 points

of the algorithm. A graphical illustration called 'Optimizer Signature' is used in the experimental approach to identify these stochastic algorithms' intrinsic bias(es). The intrinsic bias(es) of the sine cosine algorithm (SCA) are discussed below.

2.3.1 Experimental Setup

In order to obtain the idea about the intrinsic bias(es), 10 successive execution of SCA algorithm are performed on the constant objective function $f(x_1, x_2) = 1$ in the range $[-1, 1]$. In each execution, 1000 points have been generated, which means in 10 execution, 10,000 points are generated. All the points in the search space are graphically illustrated using a scatter plot to get the signature of the SCA algorithm. It is interesting to note that signature of any meta-heuristic algorithm may change upon executing the experiment several times, but the bias(es) of an algorithm is identifiable. In Figs. 2.5 and 2.6, two representative signatures of the SCA algorithm with 10,000 points are illustrated. Both of the figures may have slight differences from each other, but the patterns for intrinsic bias(es) are identifiable.

It is evident from the illustration of signature that SCA is majorly central biased and axial biased, and partially edge biased algorithm. The depiction of the signature indicates that the performance of the SCA algorithm will be badly affected if the true optima of the objective function lie in the second and fourth quadrants of the search space. On the other hand, SCA will perform better on the objective functions whose true optimum lies in the central or axial region of the search space. Further research is required to understand this biased behavior and possible modifications to eliminate the same.

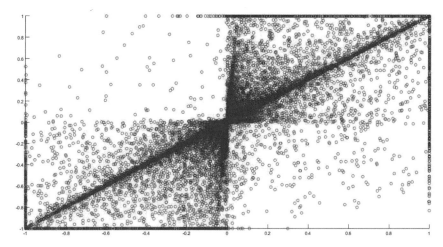

Fig. 2.6 Experiment 2—signature of SCA with 10,000 points

2.4 Numerical Example

In this section, a simple numerical example is taken to demonstrate the step-by-step working procedure of the SCA algorithm. For the sake of simplicity, the two-variable sphere function (2.8) in the range $[-5, 5]$ is considered as the underlying objective function and the optimization problem is formulated as of minimization type.

$$\text{Min } f(X) = X_1^2 + X_2^2 \text{ s.t} \tag{2.8}$$
$$X = (X_1, X_2); \ X_1, X_2 \in [-5, 5] \tag{2.9}$$

The sphere function is a simple 2-variable problem with the global minima situated at $(0, 0)$. For a simple demonstration of the computational procedure involved in the SCA algorithm, a small population size of 5 is taken, and the hand calculation for 2 iterations is added. As a first step, the population is randomly initialized in the range $[-5, 5]$ using Eq. (2.1), and the fitness values of search agents are calculated. The fitness of an individual solution is usually defined as the value of the objective function corresponding to it. Substituting $X_{1,1} = -0.6126$ and $X_{1,2} = -0.1024$ in the objective function $f = X_{1,1}^2 + X_{1,2}^2$, we get 0.3857. Similarly, we will calculate the objective function value for all other search agents (see Table 2.1).

Better objective function values represent better solutions. In this example, a solution or search agent with the least objective function value is regarded as the best solution. As one can observe from Table 2.1, the minimum objective function value is 0.3857, and therefore, $(-0.6126, -0.1024)$ is the best solution or the destination point (shown in bold). Now, the main loop of the algorithm starts, and the iteration counter (t) is initialized, $t = 0$.

Table 2.1 Initial population

Agent No.	X_{i1}	X_{i2}	Fitness value
1	−0.6126	−0.1024	0.3857
2	−1.1844	−0.5441	1.6989
3	2.6552	1.4631	9.1907
4	2.9520	2.0936	13.0977
5	−3.1313	2.5469	16.2914

First Iteration

The destination point is (−0.6126, −0.1024).

The destination fitness is **0.3857**.

Updating first search agent (i = 1)[1]

Consider the first search agent X_1, and its first component $X_{1,1} = -0.6126$ is updated. To update $X_{1,1}$, we need r_1, and it is calculated using Eq. (2.4), while r_2, and r_3 are generated randomly using Eqs. (2.5) and (2.6), respectively. Consider $r_1 = 2, r_2 = 1.7343, r_3 = 1.3594,$ and $r_4 = 0.6551$.[2]

$$X_{1,1}^1 = (-0.6126) + 2 \times \cos(1.7343)$$
$$\times |1.3594 \times (-0.6126) - (-0.6126)| = -0.6842$$

Since the updated position of $X_{1,1}$ lies in the range [−5, 5], we will accept the update. Similarly, we will update the second component $X_{1,2} = -0.1024$ by considering $r_2 = 1.0217, r_3 = 0.2380,$ and $r_4 = 0.49840$ as follows;

$$X_{1,2}^1 = (-0.1024) + 2 \times \sin(6.1720)$$
$$\times |1.5381 \times (-0.1024) - (-0.1024)| = 0.0307$$

The updated value of $X_{1,2}$ is also within the search space [−5, 5]. Thus, the updated position of the first search agent is $X_1 = (-0.6842, 0.0307)$. A similar process is used to update the all other search agents.

Updating second search agent (i = 2)

(first component) (j = 1)

Consider $r_2 = 6.0302, r_3 = 0.6808,$ and $r_4 = 0.5853$

$$X_{2,1}^1 = (-1.1844) + 2 \times \cos(6.0302)$$
$$\times |0.6808 \times (-0.6126) - (-1.1844)| = 0.3016$$

---

[1] Note that all calculations are carried out component wise.

[2] All random numbers are generated using MATLAB rand function.

(second component) ($j = 2$)
$r_2 = 1.4063$, $r_3 = 1.5025$, and $r_4 = 0.2551$

$$X^1_{2,2} = (-0.5441) + 2 \times \sin(1.4063)$$
$$\times |1.5025 \times (-0.1024) - (-0.5441)| = 0.2260$$

Updating third search agent ($i = 3$)
(first component) ($j = 1$)
$r_2 = 3.1790$, $r_3 = 1.3982$, and $r_4 = 0.8909$

$$X^1_{3,1} = (2.6552) + 2 \times \cos(3.1790)$$
$$\times |1.3982 \times (-0.6126) - (2.6552)| = -1.8088$$

(second component) ($j = 2$)
$r_2 = 6.0274$, $r_3 = 1.0944$, and $r_4 = 0.1386$

$$X^1_{3,2} = (1.4631) + 2 \times \sin(6.0274) \times |1.0944 \times (-0.1024) - (1.4631)| = 0.8479$$

Updating fourth search agent ($i = 4$)
(first component) ($j = 1$)
$r_2 = 0.9380$, $r_3 = 0.5150$, and $r_4 = 0.8407$

$$X^1_{4,1} = (2.9520) + 2 \times \cos(0.9380) \times |0.5150 \times (-0.6126) - (2.9520)| = 6.2597$$

The updated position is 6.2597, which is out of the search space. Therefore the position is set as $X^1_{4,1} = 5$ because the updated value is near to 5, the upper bound of the search space
(second component) ($j = 2$)
$r_2 = 1.5977$, $r_3 = 1.6286$, and $r_4 = 0.2435$

$$X^1_{4,2} = (2.0936) + 2 \times \sin(1.5977) \times |1.6286 \times (-0.1024) - (2.0936)| = 5.5436$$

Again the updated position is out of the search space. Therefore, the updated position is set as $X^1_{4,2} = 5$.

Updating fifth search agent ($i = 5$)
(first component) ($j = 1$)
$r_2 = 5.8387$, $r_3 = 0.7000$, and $r_4 = 0.1966$

$$X^1_{5,1} = (-3.1313) + 2 \times \sin(5.8387)$$
$$\times |0.7000 \times (-0.6126) - (-3.1313)| = -6.0054$$

Table 2.2 Updated position of the search agents

Agent No.	X_{i1}	X_{i2}	Fitness value
1	−0.6842	0.0307	0.4691
2	**0.3016**	**0.2260**	**0.1420**
3	−1.8088	0.8479	3.9907
4	5.0000	5.0000	50
5	−5.0000	5.0000	50

The updated component position is −6.0054, which is out of the search space. Therefore, the updated position is set as $X_{5,1}^1 = -5$ because the lower bound of the search space is −5

(second component) ($j = 2$)

$r_2 = 1.5776, r_3 = 1.2321, r_4 = 0.4733$

$$X_{5,2}^1 = (2.5469) + 2 \times \sin(1.5776) \times |1.2321 \times (-0.1024) - (2.5469)| = 7.0836$$

Again the updated component position is out of the search space. Therefore, the updated position is set as $X_{5,2}^1 = 5$. Finally, updated population after first iteration.

Now, termination criteria is checked. Since we planned to run the algorithm for 2 iterations and till now only one iteration is complete, we will move to iteration 2.

Second Iteration

Clearly, from Table 2.2, the minimum objective function value is 0.1420, which corresponds to the second search agent. Therefore, the best solution is (0.3016, 0.2260) and the best fitness is 0.1420.

Updating first search agent ($i = 1$)

$r_1 = 1$

(first component) ($j = 1$)

$r_2 = 2.2095, r_3 = 1.6617, r_4 = 0.5853$

$$X_{1,1}^2 = (-0.0284) + 1 \times \cos(2.2095)$$
$$\times |1.6617 \times (0.3016) - (-0.0284)| = -1.3909$$

(second component) ($j = 2$)

$r_2 = 3.4540, r_3 = 1.8344, r_4 = 0.2858$

$$X_{1,2}^2 = (1.7795) + 1 \times \sin(3.4540) \times |1.8344 \times (0.2260) - (1.7795)| = -0.0873$$

Updating second search agent ($i = 2$)

(first component) ($j = 1$)

$r_2 = 5.8678, r_3 = 1.1504, r_4 = 0.1178$

$$X_{2,1}^2 = (0.3016) + 1 \times \sin(5.8678) \times |1.1504 \times (0.3016) - (0.3016)| = 0.1487$$

(second component) ($j = 2$)
$r_2 = 3.5677, r_3 = 0.1517, r_4 = 0.0540$

$$X_{2,2}^2 = (0.2260) + (1 \times \sin(3.5677)) \times |(0.1517 \times (0.2260) - 0.2260)| = 0.1468$$

Updating third search agent ($i = 3$)
(first component) ($j = 1$)
$r_2 = 3.3351, r_3 = 1.5583, r_4 = 0.9340$

$$X_{3,1}^2 = (-1.8088) + 1 \times \cos(3.3351)$$
$$\times |1.5583 \times (0.3016) - (-1.8088)| = -3.8112$$

(second component) ($j = 2$)
$r_2 = 0.8162, r_3 = 1.1376, r_4 = 0.4694$

$$X_{3,2}^2 = (0.8479) + 1 \times \sin(0.8162) \times |1.1376 \times (0.2260) - (0.8479)| = 1.3441$$

Updating fourth search agent ($i = 4$)
(first component) ($j = 1$)
$r_2 = 0.0748, r_3 = 0.6742, r_4 = 0.1622$

$$X_{4,1}^2 = (5.0000) + 1 \times \sin(0.0748) \times |0.6742 \times (0.3016) - (5.0000)| = 5.3661$$

The updated component position is 5.3661, which is out of the search space. Therefore, the updated position is set as $X_{4,1}^2 = 5$
(second component) ($j = 2$)
$r_2 = 4.9906, r_3 = 0.6224, r_4 = 0.5285$

$$X_{2,2}^2 = (5.0000) + 1 \times \cos(4.9906) \times |0.6224 \times (0.2260) - (5.0000)| = 6.3483$$

Again the updated component position is 6.3483, which is out of the search space. Therefore, the updated position is taken as $X_{4,2}^2 = 5$.

Updating fifth search agent ($i = 5$)
(first component) ($j = 1$)
$r_2 = 1.0408, r_3 = 1.2040, r_4 = 0.2630$

$$X_{5,1}^2 = (-5.0000) + 1 \times \sin(1.0408)$$
$$\times |0.2040 \times (0.3016) - (-5.0000)| = -0.5315$$

(second component) ($j = 2$)
$r_2 = 4.1097, r_3 = 1.3784, r_4 = 0.7482$

Table 2.3 Updated positions of the search agents after iteration 2

Agent No.	x_1	x_2	Fitness value
1	−1.3909	−0.0837	1.9423
2	**0.1487**	**0.1468**	**0.0436**
3	−3.8112	1.3441	16.3318
4	5.0000	5.0000	50.000
5	−0.5315	2.2804	5.4826

$$X_{5,2}^2 = (5.0000) + 1 \times \cos(4.1097) \times |1.3784 \times (0.2260) - (5.0000)| = 2.2804$$

Finally, the updated search agents are given in Table 2.3. Now, the iteration counter is increased by one and is set to two. Since the termination criterion is met, the best solution identified by the SCA algorithm is (0.1487, 0.1468), and the optimal value of the objective function determined by the SCA algorithm is 0.0436, both of which are near to the exact solution (0, 0) and exact value 0. In the similar fashion, more iterations can be performed to further refine the obtained solution.

2.5 Source Code

In this section, the source code (2.1) of the basic SCA algorithm in MATLAB is illustrated. For simplicity and to be consistent with the numerical example presented in Sect. 2.4, the sphere function given by Eq. (2.8) is used as an objective function. The source code of the objective function which we need to minimize by using the SCA algorithm is shown in Listing 2.2.

Listing 2.1 The basic code of the SCA algorithm in MATLAB

```
1  % Sine Cosine Algorithm (SCA)
2  % MATLAB Version 2015a
3  % Reference Paper: S. Mirjalili, SCA: A Sine ...
        Cosine Algorithm for solving optimization ...
        problems
4  %   Knowledge-Based Systems, DOI: ...
        http://dx.doi.org/10.1016/j.knosys.2015.12.022
5  % Remark: This code is for academic purposes ...
        only. For any query or suggestion write to us;
6  %J.C. Bansal (jcbansal@sau.ac.in)
7  %Prathu Bajpai (prathu.bajpai1812@gmail.com)
8  %=======================%
9
10 % Initialization of Sine Cosine Algorithm
11 clc;
12 clear all;
```

```
13
14   Np = 50;      % Population size
15   Dim = 30;     % Dimension of the search space
16   Objf = @cost_function;       % Cost function or ...
         objective function
17   lb = -5.*ones(1,Dim);        % Lower bound of the ...
         search space
18   ub = 5.*ones(1,Dim);         % Upper bound of the ...
         search space
19
20   X = zeros( Np, Dim );        % Container to ...
         store population
21   fit = zeros( 1, Dim );       % Initialize ...
         fitness vector
22   T = 1000;                    % Maximum Iterations
23   t = 0;                       % Iteration counter
24
25   % Initialize parameters
26   b = 2; % Constant parameter
27   p = 0.5 % Probability switch
28
29   r1 = b; r2 = 1; r3 = 1; r4 = 1; %Initial control ...
         parameters
30
31   % Initialize population
32   for i=1:Np
33       X(i,:) = lb + rand(1,Dim).*(ub-lb);
34       fit(i) = Objf(X(i,:));
35   end
36   pop = X;       % Initial Population
37   [best_fit,ind] = min(fit);   % Best solution is ...
         destination point
38   best_agent = pop(ind,:);     % Best agent in the ...
         population
39
40   %%Iteration Loop
41
42   while t < T
43       r1 = b - t*(b/T);
44
45       %Position update equations
46       for i=1:Np
47           % Update control parameters
48           r2 = (2*pi)*rand();
49           r3 = 2*rand();
50           r4 = rand();
51
52           % Apply switch
53           if r4 < p
54               pop(i,:) = pop(i,:)+ ...
                     r1*sin(r2)*abs(r3*best_agent - ...
                     pop(i,:));   %Equation 2.2
55           else
```

```
56                    pop(i,:) = pop(i,:)+ ...
                        r1*cos(r2)*abs(r3*best_agent - ...
                        pop(i,:));     %Equation 2.3
57            end
58        end
59
60        % Check bounds
61        for i=1:Np
62            for j=1:Dim
63                    if pop(i,j) < lb(1)
64                        pop(i,j) = lb(1);
65
66                    elseif pop(i,j) > ub(1)
67                        pop(i,j) = ub(1);
68                    end
69            end
70        end
71
72        % Evaluate fitness of updated population
73        for i=1:Np
74            fit(i) = Objf(pop(i,:));
75        end
76
77        %Update the best fitness and best solution
78        [best_fit,ind] = min(fit);
79        best_agent = pop(ind,:);
80
81        %Increase iteration counter
82        t=t+1;
83    end
84
85    display(['Optimum value obtained by SCA alg. is ...
            :', num2str(best_fit)]);
86    display(['Optimum solution obtained by SCA alg. ...
            is :', num2str(best_agent)]);
```

Listing 2.2 Cost function defined in MATLAB

```
1    %Sphere Function
2    function f = cost_function(x)
3    f = sum(x.^2);
4    end
```

Practice Exercises

1. Apply SCA to solve the sphere function problem for 10, 30, 50, and 100 variables. Compare and analyze the obtained results.
2. Discuss the influence of the population size on the performance of SCA.

3. Maximum number of iterations plays an important role in ensuring quality solutions. Explain?

Reference

1. S. Mirjalili, SCA: a sine cosine algorithm for solving optimization problems. Knowl.-Based Syst. **96**, 120–133 (2016)

Chapter 3
Sine Cosine Algorithm for Multi-objective Optimization

In many real-world situations, we have to deal with multiple objectives simultaneously in order to make appropriate decisions. The presence of multiple objectives in an optimization problem makes the problem challenging because most of the time these objectives are conflicting in nature. For example, we may want to maximize the return on investment of a portfolio and, on the other hand, minimize the risk associated with the assets in the portfolio. We may want to minimize the cost of a product while maximizing the performance of that particular product. Similarly, there are situations where we may want to maximize more than one objective at a time and minimize multiple objectives for a given optimization problem. For instance, a product manager in an XYZ mobile manufacturing company is supervising the launch of a new smartphone in the market. He/she will have to consider many features and configurations of the smartphone before launching. He/she might have to consider features like the screen resolution, size of the screen, thickness of the phone, camera resolution, battery life, operating system, and even aesthetics of the product. On the other hand, he/she might also want to minimize the amount of labor, time of production, and overall cost associated with the project. He/she knows that the objectives, in this case, are conflicting, and simultaneously achieving every objective in not possible. The solution to this dilemma is to look for some trade-off solutions so that the main motive of the problem can be served.

Even if we consider a simple problem of mobile-buying decision-making problem for an individual buyer or consumer who wishes to buy a smartphone from the available set of options, he/she might have to face the same kind of dilemma as the product manager in the company XYZ was facing. The individual smartphone buyer may want to maximize the quality and features, like size, camera quality, user-interface, aesthetics, reliability and, on the other hand, try to minimize the cost. The graphical representation of the alternative solutions in the mobile-buying decision-making problem is illustrated in Fig. 3.1.

From the above discussion, we may convince ourselves that the single-objective optimization problems are not sufficient to deal with a large class of decision-making

© The Author(s) 2023
J. C. Bansal et al., *Sine Cosine Algorithm for Optimization*,
SpringerBriefs in Computational Intelligence,
https://doi.org/10.1007/978-981-19-9722-8_3

Fig. 3.1 Mobile-buying
decision-making problem

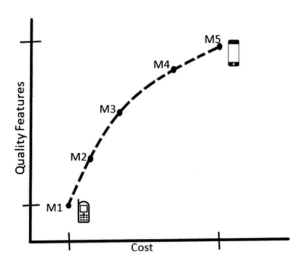

problems, where multiple objectives are present in the problem. Unlike single-objective optimization problems, the optimal solution(s) may or may not exist in multi-objective optimization Problems (MOOP). Objectives in the MOOP are conflicting in nature to each other. The optimal value of one objective may not be the optimal value for other objectives, and sometimes, it may be even worse for other objectives. For example in the above mobile-buying decision-making problem, if a buyer wants to minimize the cost for buying, and chooses the cheapest option M1, then he/she has to give up on the quality and features. Similarly, if he/she chooses the option of M5 with maximum quality and features, he/she has to loosen the pocket, and have to bear the maximum cost. In the next section, MOOP is discussed mathematically in detail.

3.1 Multi-objective Optimization Problems (MOOP)

Optimization problems involving multiple objective functions (>1), are regarded as the multi-objective optimization problems (MOOP). The underlying objective functions in MOOP can be minimization type, maximization type, or a combination of both (min–max). The procedure of finding single or multiple optimal solutions for MOOP is called multi-objective optimization.

Mathematically, a MOOP can be written in the following general form:

$$\text{Minimize } F(\bar{X}) = \left[f_1(\bar{X}), f_2(\bar{X}), \ldots, f_M(\bar{X}) \right],$$

$$\text{s.t.} \quad g_j(\bar{X}) \leq 0 \quad j = 1, 2, \ldots, J$$

$$h_k(\bar{X}) = 0 \quad k = 1, 2, \ldots, K \tag{3.1}$$

$$x_i^l \leq x_i \leq x_i^u \quad \forall i = 1, 2, \ldots D$$

where $\bar{X} = (x_1, x_2, \ldots, x_D)$ is the vector of decision variables, x_i ($i = 1, 2, \ldots D$). $g_j(\bar{X})$ and $h_k(\bar{X})$ are J inequality and K equality constraints, respectively. x_i^l and x_i^u are lower bound and upper bound constraints for decision variable x_i ($i = 1, 2, \ldots D$).

The multi-objective optimization problems give rise to two different kinds of spaces. The space $\mathcal{F} \subseteq \mathcal{R}^D$ spanned by the vectors of decision variables ($\bar{X}$) is called the decision space or search space. And, the space $\mathcal{S} \subseteq \mathcal{R}^M$ formed by all the possible values of objective functions is called the objective space or solution space [1].

Similar to single-objective optimization problems, a multi-objective optimization problem can be classified as linear and nonlinear MOOP depending on the objective functions and constraints. If all the objective functions and constraints are linear, then MOOP is referred as a linear multi-objective optimization problem. On the other hand, if any of the objectives or constraints is nonlinear, MOOP is called the nonlinear multi-objective optimization problem. Further, a MOOP can also be classified as a convex and non-convex multi-objective optimization problem. For the detailed classification, an interested reader can refer to 'Multi-Objective Optimization Using Evolutionary Algorithms' by Deb [2].

3.2 Multi-objective Optimization Techniques (MOOT)

In the mobile-buying decision-making problem, we see that our hypothetical smart buyer wants to maximize the quality and features, but also wants to minimize the cost. We cannot generalize this case for every buyer or consumer in the market, there might be some buyers not worried about the cost, and their primary preference is toward quality and features only. Similarly, there might be buyers available in the market who do not think about the quality and features, but make their decision on the basis of the cost only. In both of these extreme cases, the problem is in fact not a multi-objective but a single objective. In many situations, due to the scarcity of resources taking decisions on extremes is not a feasible option. One has to make some trade-off on available choices based on his/her preference. In that case, this problem becomes a multi-objective optimization problem.

The gist of the above discussion is, when multiple-conflicting objectives are important in the decision-making process, finding a single optimum solution such that it optimizes all the objectives simultaneously is not possible, and even not prudent to look for. We have to settle ourselves on some trade-off solutions, or in layman's language, we have to achieve certain harmony between conflicting objectives based on our preference. If in case, the harmony or balance between these conflicting objectives is not possible, we must try to find a list of preferences as to which objective should be given the most preference and make a compromise.

Multi-objective optimization techniques are methods or procedures, primarily focused on dealing with the optimization problems, where conflicting objectives cannot be ignored. There are classical and evolutionary techniques available in the literature of multi-objective optimization techniques, which we will discuss in later sections of this chapter, but before going further, we have to make ourselves familiar with some concepts and terminologies, which are important for understanding the multi-objective optimization procedures.

3.2.1 Some Concepts and Terminologies

For understanding the idea of optimality in multi-objective optimization, first we have to discuss about the Pareto optimality. The concept of Pareto optimality was first introduced by Francis Ysidro Edgeworth, and it is named after the mathematician Vilfredo Pareto, who generalized the concept for multi-objective optimization [3].

Suppose we have a minimization problem as mentioned in Eq. (3.1). The multi-objective function is denoted by $F(\bar{X}) = [f_1(\bar{X}), f_2(\bar{X}), \ldots, f_M(\bar{X})]$. $\mathcal{S}$ is the objective space. And, $\mathcal{F}$ is decision space, and $\bar{X} \in \mathcal{F}$ is a decision vector or solution. A reader should not get confused with the terms decision vectors and solutions, because in the literature of multi-objective optimization, these terms are used interchangeably, and have the same meaning. Continuing this tradition, we also use the terms decision vectors or solutions, interchangeably depending on the requirements.

Definition 3.1 (*Dominance*) A solution, $X_1 \in \mathcal{F}$ dominates another solution $X_2 \in \mathcal{F}$, if the following two conditions are satisfied:

1. $f_k(X_1) \leq f_k(X_2), \forall k = 1, 2, \ldots M$
2. There exists some $j \in \{1, \ldots M\}$ such that, $f_j(X_1) < f_j(X_2)$.

In the above definition, condition (1) says solution X_1 is no worse than solution X_2 in all the objectives, while condition (2) indicates that there exists at least one objective (say f_j) for which X_1 is strictly better than X_2. If any of the above conditions are not satisfied or violated, the solution X_1 does not dominate the solution X_2. It is also worth mentioning here again that we are considering a minimization-type MOOP for our discussion, and if the underlying MOOP is maximization type, the inequalities ($\leq$, $<$) will be replaced by ($\geq$, $>$) in the above definition of dominance.

If solution X_1 dominates the solution X_2, we can denote this situation mathematically as $X_1 \prec X_2$. Apart from saying, solution X_1 dominates the solution X_2, one can also say, solution X_2 is dominated by solution X_1, or X_1 is non-dominated by X_2, or solution X_1 is non-inferior to solution X_2. The concept of dominance is graphically shown in Fig. 3.2.

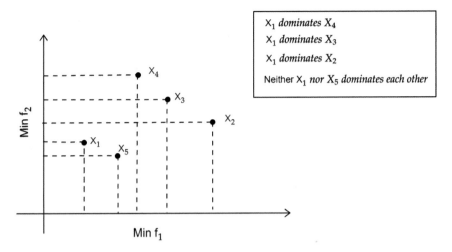

Fig. 3.2 Graphical illustration of dominance

Definition 3.2 (*Pareto optimality*) A decision vector or solution, $X^* \in \mathcal{F}$ is called Pareto optimal solution or non-dominated solution if $\nexists$ any $k \in \{1, \ldots M\}$ such that $f_k(X) < f_k(X^*)$.

The Pareto optimality of solutions implies that there does not exist any feasible solution in the decision space which would decrease some objectives without simultaneously causing an increase in at least one objective. That is, any improvement in one objective results in the worsening of at least one other objective.

Definition 3.3 (*Pareto optimal set*) The set containing all the Pareto optimal solutions is called the Pareto optimal set, P^*. It is given by,

$$P^* = \{X^* \in \mathcal{F} : X^* \prec X, \ \forall X \in \mathcal{F}\} \tag{3.2}$$

Definition 3.4 (*Pareto optimal front*) In the objective space $\mathcal{S}$, all the objective values corresponding to the Pareto optimal solutions are joined with a continuous curve. This curve is called the Pareto optimal frontier, simply Pareto optimal front.

The graphical illustration of the Pareto front and Pareto optimal set is shown in Fig. 3.3. For every solution in the decision space, there is a corresponding objective value in the objective space. The objectives of the optimization problem are to be minimized, and these objectives are conflicting in nature. Furthermore, a vector is called an ideal vector or Utopian objective vector if it contains all the decision variables that correspond the optima of objectives functions when each objective is considered separately [1]. It is also interesting to mention that when the objective functions in a MOOP are not conflicting in nature, the cardinality of the Pareto optimal set is one [5]. In the next section, different approaches to handle multi-objective optimization problems are discussed in detail.

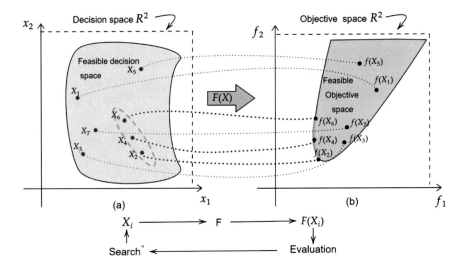

Fig. 3.3 Graphical illustration of mapping of a decision space onto an objective space, where both the conflicting objectives are to be minimized [4]

3.2.2 Different Approaches of Solving MOOP

The conflicting objectives in the multi-objective optimization (MOO) problems lead to multiple trade-off solutions or Pareto optimal solutions. Many different approaches to solve MOOP are proposed and classified for the MOO problems. There are conventional or classical methods and modern or meta-heuristic methods available in the literature. In conventional methods, the reformulation of the MOO problems is required to proceed with the optimization process. Different methods and approaches were proposed to reformulate the MOO problems. For instance, one approach is to reformulate MOO problem into single-objective optimization problem using weighted sum of objective functions in which weights are assigned on the basis of a preference or utility by decision-maker (DM). One other approach is to optimize the most preferred objective function of DM's interest and treat other objectives as constraints with some predefined limits. In both of these methods, some preference of the decision-maker is required before the optimization process begins. However, some classical methods do not need any priori information about the relative importance of the objective function. These methods are called 'No-preference methods'. Discussing these classical methods in detail is not in the scope of this book. For detailed information regarding classical methods, an interested reader can refer to the book, "Multi-Objective Optimization Using Evolutionary Algorithms" by Deb [2]. How-

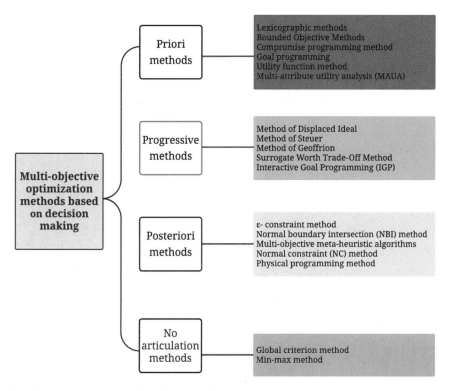

Fig. 3.4 Classification of multi-objective optimization methods

ever, for the convenience of the readers, a brief classification inspired by Miettinen [6], and Hwang and Masud [7] is presented here (Fig. 3.4).

(1) **Priori methods**: In priori methods, the preference information (e.g., weights of the objective functions) is specified before applying the optimization algorithm. These preferences are used to quantify the relative importance of the different objective functions in the MOO problems. These methods convert a MOO problem into a single-objective optimization problem for the further optimization process. Priori approaches can be described as a "decide first and then search" approaches, where the decision is taken before searching. The major limitation of these methods is that they are applicable only when the decision-maker knows the problem very well. However, it is very challenging for the decision-maker to accurately express his/her preferences through some goals or weights. Moreover, every time the relative importance of the objectives changes, weights and preferences are to be relooked. Some example of priori methods are bounded objective method, lexicographic method, compromise programming method, goal programming, utility function method, and multi-attribute utility analysis (MAUA) [7].

(2) **Progressive (or Interactive) methods**: In progressive or interactive methods, to guide the search process, the objective functions and constraints are redefined and incorporated multiple times, based on the decision-maker's preferences, during the execution of the algorithms. A subset of non-dominated (Pareto optimal) solutions are found in each iteration, and the resultant Pareto set is then presented before the decision-maker. If the decision-maker is satisfied with the solutions, then the algorithm terminates the optimization process. However, if decision-maker is not satisfied, then he/she is required to modify the preferences, and new Pareto optimal solutions are found using the new modified preferences. This process continues until the decision-maker is satisfied or no further improvement is possible. Method of displaced ideal, method of Steuer, method of Geoffri on, interactive goal programming (IGP), and surrogate worth trade-off method are some of the methods that fall under this category [7].

(3) **"A Posterior" approach**: These approaches are mainly 'first search and then decide' strategies, where the search is executed before decision-making. The non-dominated solutions are first generated using some optimization method. Once the method is terminated, the most satisfactory solutions are selected from the obtained non-dominated solutions based on the decision-maker's requirements. In other words, the decision-maker expresses his/her preferences once all the non-dominated solutions are generated. The decision-making process is involved after the solutions are generated, with the changing preferences of the decision-maker, new decisions are possible without repeating the optimization process. The main criticism about the posterior approaches is that these methods usually generate many non-dominated solutions, making it very difficult for the decision-maker to choose the most satisfactory solution. Moreover, the process of approximating the Pareto optimal set is often time-consuming. Some of the examples are ϵ-constraint method, physical programming method, normal boundary intersection (NBI) method, and normal constraint (NC) method [7].

(4) **No articulation approach**: In these methods, personal preference information from the decision-maker is not needed once the problem is formulated; i.e., constraints and objectives are defined. These approaches are advantageous for problems where the decision-maker cannot precisely define his/her preferences. These methods are used only when the decision-maker is not available, or the decision-maker cannot define what he/she prefers. These methods are known for their faster convergence and speed. Some examples of these methods are the global criterion method and the min–max method [7].

The approaches discussed above often lead to a solution that may not be optimal. The obtained Pareto front might be locally non-dominated, not necessarily globally non-dominated. For example, approaches in which a multi-objective optimization problem is reformulated as single-objective optimization problems. The reformulation is sometimes challenging. Also, converting the objectives into constraints due to the conflicting nature of the multiple objectives is not feasible. Similarly, in the weighted sum approaches, the major challenge is to determine the appropriate weights based on the preference of the user. Many complex real-world problems do not provide sufficient information about the problem, and hence, it is not an easy task to get the optimal values of the weights. Moreover, in most of the methods mentioned above, additional parameter settings are required. Decision-maker is supposed to supply the value of parameters, and the preferences of the decision-maker are subjective in many cases. These methods are not only difficult to implement, but they also suffer from many drawbacks. Some are mentioned below:

(1) Most of these methods fail to perform if the shape of the Pareto front is concave or disconnected.
(2) These methods are only able to produce a single solution in every run of the optimization process. For obtaining different trade-off solutions, one has to run the algorithm multiple times, which increases the computational cost of these methods.
(3) The different objectives might take values of different orders of magnitude (or different units). A normalization of objective functions is required, which demands knowledge of the extremum values of each objective in the objective space.

The methods for multi-objective optimization are presented above utilize the single-objective optimization techniques for the optimization process. The single-objective optimization techniques are incapable of producing multiple solutions, which is the most important aspect of the MOO problems.

The challenge of producing multiple solutions for a MOO problem, however, can be handled in a more sophisticated manner. There are other promising methods available, which are non-conventional, more advanced, and intelligent. The methods which require very low (or, no) information about the optimization problems and are equipped with the potential of producing multiple solutions in a single run of the optimization process. Moreover, they provide privilege to the user in deciding the number of solutions, as much as he wants, or as low as he can. These methods are meta-heuristic methods. The population-based approach and capability of handling black-box problems make these evolutionary and swarm-based techniques a suitable candidate for MOO problems. Meta-heuristic techniques for single-objective optimization can be extended to handle MOO problems with some modifications, because of their basic structure, which is different from the single-objective optimization.

The first hint regarding the possibility of using population-based stochastic optimization algorithms to solve multi-objective optimization problems was presented in the Ph.D. thesis of Rosenberg [8], in which a multi-objective problem was restated as a single-objective problem and solved with the genetic algorithm (GA). However, David Schaffer was the first who introduced the revolutionary idea of applying stochastic techniques to deal with multi-objective optimization problems by proposing the multi-objective evolutionary optimization approach based on the genetic algorithm (GA), known as vector evaluated genetic algorithm (VEGA) [9]. The expansion in the research of meta-heuristic techniques and advancements in the computing power of modern computers paved the way for the researchers to focus on articulating more superior multi-objective meta-heuristic techniques. For example, some of the well-known multi-objective stochastic optimization techniques are non-dominated sorting genetic algorithm (NSGA) [10], non-dominated sorting genetic algorithm version 2 (NSGA-II) [11], multi-objective particle swarm optimization (MOPSO) [12], Pareto archived evolution strategy (PAES) [13], Pareto-frontier differential evolution (PDE) [14], multi-objective ant colony optimization [15], multi-objective dragonfly algorithm (MODA) [16], and multi-objective sine cosine algorithm [17].

The population-based approach of meta-heuristic algorithms provides liberty to obtain multiple Pareto optimal solutions in a single run of the algorithm. Instead of finding a single Pareto optimal front containing solutions with specific preferences, these methods explore the search space extensively to provide multiple Pareto optimal front corresponding to the different regions. In the next section, we will discuss the particular case of multi-objective sine cosine algorithm, which is the main focus of this chapter.

3.3 Multi-objective SCA

The basic structure of multi-objective optimization is different from the single-objective optimization, which compels to incorporate some modifications in the original sine cosine algorithm (SCA) proposed for single-objective optimization. Before coming to the proposed modifications in the SCA, let us discuss some problems, which have to be taken into consideration.

1. **How to choose P_g (i.e., destination point) in each iteration?**
   SCA is required to favor non-dominated solutions over dominated solutions, and drive the population toward the different parts of the Pareto front, or set of non-dominated solutions, and not only in the direction of the destination point.
2. **How to identify the non-dominated solutions in SCA, and how to retain the solutions during the search process?**
   Ans: One strategy is to combine all solutions obtained in the optimization process and then extract the non-dominated solutions from the combined population. Of course, other approaches do exist.
3. **How to maintain the diversity in the population, so that a set of well-distributed non-dominated solutions can be found along the Pareto front?**
   Ans: Some classical niching methods (e.g., crowding or sharing) are available and can be adopted for maintaining the diversity.

The problem of finding an accurate approximation of the true Pareto optimal front is challenging and even sometimes impossible for a given multi-objective optimization problem. However, the approximated Pareto front obtained by using multi-objective meta-heuristic algorithms should possess certain characteristics. For instance, the resultant non-dominated set of solutions should lie at a minimum distance from the optimal Pareto front and the solutions in the resultant Pareto front should be uniformly distributed to cover a wide range of the non-dominated solutions [18]. These points were taken into consideration, and various attempts have been made to design the multi-objective SCA. The structure of multi-objective SCA is different because of the presence of Pareto optimal solutions and the concept of domi-

Table 3.1 Multi-objective sine cosine algorithms

Approach	Algorithm name	Fitness assignment	Diversity mechanism	Elitism	External population	Selection of fittest solution
Non-dominance and diversity based	MOSCA [17]	Ranking based on non-domination sorting	Crowding distance	Yes	No	Crowded comparison operator
	MOSCA by Raut et al. [19]	Ranking based on non-domination sorting	Crowding distance	Yes	No	Fuzzy decision-making
	MOSCA_SSC [20]	Ranking based on non-domination sorting	Crowding distance	Yes	No	Knee-point based
	MOSCA by Selim et al. [21]	Ranking based on non-domination sorting	Grid mechanism and leader selection mechanism	Yes	Yes	Fuzzy logic decision-making
	MOSCA with fuzzy loss sensitivity factor (FLSF) [22]	Ranking based on non-domination sorting	Grid mechanism and leader selection mechanism	Yes	Yes	Grey relational analysis
	MOCSCA [23]	Ranking based on non-domination sorting	Grid mechanism and leader selection mechanism	Yes	Yes	Grey relational analysis
Aggregation based	SCA for CEED [24]	Aggregated objective function using max–max price penalty factors	No	No	No	Based on fitness value
	Multi-objective ISCA [25]	Fuzzy membership function	No	No	No	Based on fitness value
Mixed	MSCO [26]	Weighted average of normalized objective	Randomly assigned weights + opposition-based learning strategy	Yes	Yes	User selection
	NSCA [27]	Weighted average of normalized objective	Randomly assigned weights	Yes	Yes	User selection

nance; however, the search mechanism is almost similar to single-objective SCA. We will study the multi-objective versions of SCA based on two approaches, particularly the aggregation-based approaches and non-dominance diversity-based approaches which are discussed in subsequent sections. A list of all the multi-objective SCA proposed in the literature is presented in Table 3.1.

3.3.1 Aggregation-Based Multi-objective Sine Cosine Algorithm and Their Applications

In aggregation-based approaches, the multiple objectives of a MOO problem are combined using aggregation operator to form a single-objective function. The aggregation operators merge multiple objectives using techniques like random weights, price penalty function, fuzzy membership function, utility function, etc. [25] to formulate a single-objective function. This single objective is then solved using standard single-objective optimization algorithms. However, in principle, aggregation-based approaches for handling MOO problem fail to find solutions when the Pareto optimal region is non-convex. Fortunately, not many real-world multi-objective optimization problems have been found to have a non-convex Pareto optimal region. This is the reason why the aggregation-based approaches are still popular and used in practice for multi-objective optimization problems [2].

The single-objective sine cosine algorithm [28] is a robust optimizer and can be utilized with aggregation-based approaches for solving MOO problems. Some significant applications of aggregation-based MOO-SCA are discussed here in the subsequent sections.

3.3.1.1 Multi-objective Improved Sine Cosine Algorithm for Optimal Allocation of STATCOM

In power systems, STATCOM or static synchronous compensator is a power electronic device used in power systems to regulate its various parameters either by injecting or by absorbing the reactive power. The optimal location of STATCOM is needed to enhance the performance of the power system and simultaneously reduce the cost. Multi-objective improved SCA was proposed by Singh and Tiwari [25] to handle the problem of optimal allocation of holomorphic embedded load-flow (HELF) model of STATCOMs with six objective functions. In the proposed improved SCA (ISCA), some modifications were incorporated in the SCA to boost its exploration and exploitation capabilities. The control parameter r_1 is modified to change the range of sine and cosine functions in an adaptive manner.

$$r_1 = \gamma \times \cos\left(90° - 90°\left(\frac{t-T}{T}\right)\right) \times \cos\left(60° - 60°\left(\frac{t-T}{T}\right)\right) \qquad (3.3)$$

where γ is a constant and its value is taken equal to 2.

Singh and Tiwari [25] formulated STATCOM's multi-objective problem into a single-objective problem using aggregation-based approach. The underlying aggregation operator was based on a fuzzy membership function. Multiple objectives were aggregated using the concept of the fuzzy membership function. In fuzzy membership, each objective function was assigned a membership value, and these membership values represent the weights of the objectives in the aggregated fuzzy membership function. The range of fuzzy membership values lies in the interval [0, 1]. The membership value 0 indicates the incompatibility of an objective function with the aggregated function, meanwhile, the membership value 1 represents the complete compatibility of an objective function with aggregated function [29].

The underlying six objectives of the holomorphic embedded load-flow (HELF) model for STATCOMs problem can be considered as different important factors to consider before planning and operation of STATCOMs allocation. All the six objectives, (say) $f_1, f_2, f_3, f_4, f_5, f_6$, are minimization-type problems and share relative importance in the STATCOMs allocation problem. For instance, f_1 represents active power loss, f_2: reactive power loss, f_3: node voltage deviation, f_4: cost of STATCOM, f_5: node severity to voltage collapse, and f_6 denotes the apparent power flows through transmission lines. For the mathematical definition of the mentioned objectives, readers can refer to Singh and Tiwari [25].

The objective function f_3, the node voltage deviation, is a important metric for the STATCOMs allocation problem. The authors used an exponential membership function mentioned in Eq. (3.4) to compute the membership value for the objective function f_3. The exponential membership function helps in detecting good and bad solutions of the node voltage profile by assigning higher membership values to the better solutions and low membership values to other solutions. The membership value of the rest objective functions f_i, $(i = 1, 2, 4, 5, 6)$ was calculated using an quarter cosine membership function (μf_i) given by Eq. (3.5). The quarter cosine membership function help to retain the solutions of moderate quality as well, along with the solutions of high quality.

$$\mu f_1 = \begin{cases} 1 & \text{if } f_{1,\min} \leq f_1 \leq f_{1,\max} \\ e^{m \times |1 - V_k|} & \text{if } f_{1,\min} \geq f_1 \geq f_{1,\max} \end{cases} \tag{3.4}$$

where V_k is the kth bus voltage, and $m = -10$ is used to vary the time constant of an exponential curve.

$$\mu f_i = \begin{cases} 1 & \text{if } f_i \leq f_{i,\min} \\ \cos\left[\frac{\pi}{2} \times \left(\frac{f_i - f_{i,\min}}{f_{i,\max} - f_{i,\min}}\right)\right] & \text{if } f_{i,\min} < f_{i,\max} \\ 0 & \text{if } f_i \geq f_{i,\max} \end{cases} \tag{3.5}$$

where μf_i is the value of the membership function for the objective f_i, while $f_{i,\min}$ and $f_{i,\max}$ are lower and upper bounds of the ith objective.

The fuzzy membership functions μf_i of the objectives functions f_i were aggregated to produce trade-off solutions. To aggregate these fuzzy membership functions, the 'max-geometric mean' operator is used [30]. Max-geometric mean operator first calculates the geometric mean of fuzzy membership functions of the underlying objective functions as mentioned in Eq. (3.6) below,

$$\mu f = (\mu f_1 * \mu f_2 * \mu f_3 * \mu f_4 * \mu f_5 * \mu f_6)^{(\frac{1}{6})} \tag{3.6}$$

The geometric mean of the aggregated fuzzy membership function, denoted by μf, represents the degree of overall fuzzy satisfaction that means μf indicates the relative importance of the fuzzy membership functions in the aggregation. In the second procedure of the max-geometric mean operator, μf with a maximum degree is considered to generate the best trade-off solutions [25]. The given multi-objective optimization problem was reformulated as the minimization problem, mentioned below in Eq. (3.7).

$$\min f = \frac{1}{1 + \mu f} \tag{3.7}$$

f is selected as the fitness function in the proposed ISCA [25], and it provides the optimal solution without violating any of the constraints of the given multi-objective optimization problem.

3.3.1.2 Multi-objective Sine Cosine Algorithm for Combined Economic Emission Dispatch Problem

Combined economic and emission dispatch (CEED) is the process of determining the outputs of generating units in a power system in order to minimize the fuel cost and pollutants emissions at the same time. Gonidakis and Vlachos [24] solved the combined economic emission dispatch (CEED) problem using sine cosine algorithm (SCA). The objective of the CEED problem is to minimize the four conflicting objective functions of fuel cost, nitrogen oxides (NO_x) emission, sulfur dioxide (SO_2) emission, and carbon dioxide (CO_2) emission, under certain constraints. This multi-objective problem is converted into a single objective by introducing penalty factors to the objectives representing pollutants [31]. Moreover, to deal with the constraints, penalty function method is used. The authors used the max–max price penalty factor to solve the CEED problem, which is the ratio between maximum fuel cost and maximum emission of the corresponding generator [31]. It is expressed in Eq. (3.8).

$$h_i = \frac{F(P_{i,\max})}{E(P_{i,\max})}, \quad i = 1, 2, \ldots n \tag{3.8}$$

where $F(P_{i,\max})$ is the maximum fuel cost, $E(P_{i,\max})$ is the maximum emission, n is the number of generating units, and P_i is the active power generated by the ith generating unit.

In real-time economic emission dispatch, generator fuel cost curves are approximated using polynomials. This is a standard practice followed by the industries, and this approximation greatly affects the accuracy of the economic dispatch solutions. Fuel cost and emission are usually formulated as a second-order polynomial (quadratic) functions. However, by introducing higher order polynomials, economic emission dispatch solutions can be improved. Higher order polynomial models replicate the actual thermal generators' fuel and emission costs. Gonidakis and Vlachos [24] used cubic polynomials to express the economic and emission cost. The CEED problem is mathematically formulated as mentioned in Eq. (3.9) below.

$$\text{Min } C = \sum_{i=1}^{n} [F(P_i) + h_{\text{SO}_2,i} E_{\text{SO}_2} + h_{\text{CO}_2,i} E_{\text{CO}_2} + h_{\text{NO}_x,i} E_{\text{NO}_x}]$$

$$\text{subject to } \sum_{i=1}^{n} P_i - P_\text{D} - P_\text{L} = 0$$

(3.9)

where $h_{\text{SO}_2,i}$, $h_{\text{NO}_x,i}$, $h_{\text{CO}_2,i}$ are the penalty factors of SO_2, NO_x, CO_2 emissions, respectively. E_{SO_2}, E_{CO_2}, and E_{NO_x} are the total SO_2, CO_2, and NO_x emissions, respectively. $\sum_{i=1}^{n} P_i$ is the total output of all generating units, P_D is power system load demand and P_L is the transmission loss. The constraint mentioned in Eq. (3.9) is known as the power balance constraint.

To satisfy the equality constraint, the objective function in the CEED problem is modified as follows:

$$\text{Min } G = C + k \left| \sum_{i=1}^{n} P_i - P_\text{D} - P_\text{L} \right|$$

(3.10)

where k is a constant penalty parameter.

3.3.2 Non-dominance Diversity-Based Multi-objective SCA and Its Applications

The non-dominance diversity-based approaches do not reformulate a multi-objective optimization problem into single-objective optimization problem. All the objectives are considered at the same time during the optimization process, and no preferences or weights are required to proceed. These methods produce a set of non-dominated solutions distributed uniformly along the Pareto optimal front. In the non-dominance diversity-based approaches, the very first task of the algorithm is to find non-dominated set of solutions from a given set of solutions. Different methods and procedures are available in the literature for this purpose, for example 'Naive and Slow' approach, 'continuously updated' method, and 'non-dominated sorting'

[5]. For a detailed discussion of these methods, an interested reader can refer to the book 'Multi-Objective Optimization Using Evolutionary Algorithms' by Deb [5].

The other important task in non-dominance diversity-based approaches is to maintain the distribution of non-dominated solutions throughout the Pareto region, and it is an important assessment metric for such algorithms. There are several methods for maintaining diversity, such as the adaptive grid mechanism [2], and the crowding distance mechanism [11]. These mechanisms consist of a procedure that divides objective space in a recursive manner. Next, we will discuss about the first multi-objective version of SCA based on non-dominance diversity approach.

3.3.2.1   Multi-objective Sine Cosine Algorithm (MOSCA)

Tawhid and Savsani [17] proposed the first multi-objective version of SCA using elitism-based non-dominated sorting and crowding distance (CD) method of NSGA-II [11]. In MOSCA, the elitist non-dominated sorting adopted to introduce the selection bias to the solutions (or, agents) in the population, enabling the model to select the solutions from the fronts closer to the true Pareto optimal front (let us denote Pareto optimal front by PF^*). To maintain the diversity in the population, the crowded-comparison approach of NSGA-II was adopted. The working of MOSCA can be divided into two phases:

1. Elitist non-dominated sorting.
2. Crowding distance assignment and comparison.

Elitist non-dominated sorting
In elitist non-dominated sorting approach, for each solution, two attributes are defined:

(i) domination count (n_i): number of solutions dominating the solution X_i
(ii) S_i, a set of solutions dominated by the solution X_i, are calculated using Procedure 1.

All the solutions X_i that are assigned a domination count $(n_i = 0)$, are put in the first non-dominated level (or, first Pareto front) (PF_1), and their non-domination rank (NDR_i) is set equal to 1 (see Procedure 1). Then, for obtaining the second non-domination level for each solution X_i with $n_i = 0$, each member X_j of the set S_i is visited, and its domination count n_j is reduced by one. While reducing domination count if it falls to '0', the corresponding solution X_j is put in the second non-domination level (PF_2), and its rank (NDR_j) is set equal to 2. The above procedure is repeated for each member of the second non-domination level to identify the third non-domination level. This process continues until the whole population is classified into different non-domination levels (see Procedure 2).

Procedure 1: Determining the optimal non-dominated set
Step 1 For each $X_i \in P$ (Population), $i \in 1, 2, \ldots N$, set $n_i = 0$ and $S_i = \phi$.
Then set solution counter $i = 1$.
Step 2 for all $j \in \{1, 2, \ldots N\}$ and $j \neq i$, If $X_i \prec X_j$, update $S_i = S_i \cup X_j$.
Otherwise, if $X_j \prec X_i$, set $n_i = n_i + 1$.
Step 3 Replace i by $i + 1$. If $i \leq N$, go to step 2. Otherwise, go to step 4.
Step 4 Keep X_i in P_1 (first non-dominated front) if $n_i = 0$ and set $\text{NDR}_i = 1$.

Procedure 2: Non-dominated sorting
Step 1 Determine the best non-dominated set or front (P1) using procedure 1
Step 2 Set a front counter$(k) = 1$
Step 3 While $P_k \neq \phi$, perform the following steps
Step 3(a) Initialize $Q = \phi$ for storing next non-dominated solutions
Step 3(b) For each $X_i \in P_k$ and for each $X_j \in S_i$, update $n_j = n_{j-1}$
Step 3(c) If $n_j = 0$, keep X_j in Q (i.e., $Q = Q \cup \{X_j\}$ and set $\text{NDR}_j = k + 1$
Step 4 Set $k = k + 1$ and $P_k = Q$, go to Step 2.

Crowding distance estimation

For measuring the distribution of the solutions in the neighborhood of a solution, MOSCA adopted the crowding distance metric as used in the NSGA-II [11]. The crowding distance metric estimates the normalized search space around a solution X_i which is not occupied by any other solution in the population. The crowding distance value of a particular solution is the average distance of its two neighboring solutions. The crowding distance is calculated by sorting all the solutions in the population of a particular non-dominated set in ascending order for each objective function f_l $(l = 1, 2 \ldots M)$. The individuals with the lowest and the highest objective function values are assigned an infinite crowding distance so that they are always selected, while other solutions are assigned the crowding distance (cd_l^i) using the following equation:

$$\text{cd}_l^i = \frac{f_l^{i+1} - f_l^{i-1}}{f_l^{\max} - f_l^{\min}} \quad \forall l = 1, 2 \ldots M, i = 2, 3 \ldots (l - 1) \tag{3.11}$$

The final crowding distance value (CD_i) for each solution $(X_i, i = 1 \ldots N)$ is computed by adding the solution's crowding distance values (cd_l^i) in each objective function.

$$\text{CD}_i = \sum_{l=1}^{m} \text{cd}_l^i \tag{3.12}$$

For $m = 2$, the crowding distances of a set of mutually non-dominated points are illustrated in Fig. 3.5.

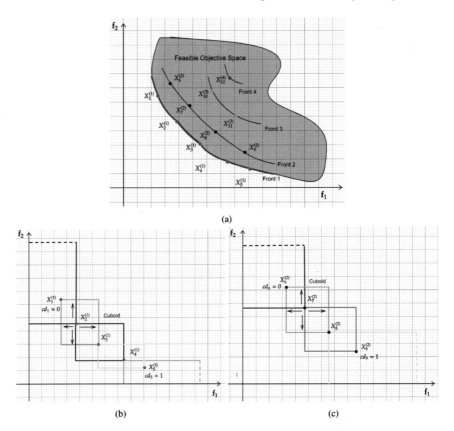

(a)

(b)   (c)

Fig. 3.5 Non-dominance ranking and crowding distance

Crowded tournament selection

After calculating the crowding distance (CD) for each of the solutions (see Procedure 3), the SCA is operated to generate a new population P_j. The new and the old population (P_o) are then merged to form a population P_{new} of size greater than N. In order to maintain a constant population size N, a crowded tournament selection operator (defined in 3.5) based on the non-dominated ranking (NDR), and the crowding distance (CD) are used to select N solutions from the P_k number of solutions to form the updated population.

Definition 3.5 (*Crowded tournament selection operator*) A solution X_i is selected over solution X_j if it satisfies any of the following conditions: 1. If solution X_i has a lower (or, better) NDR than X_j, 2. If solutions have the same NDR but solution X_i has a better crowding distance (CD) than the solution X_j.

That means, between solutions with different NDRs, we prefer the solutions with the lower rank. And, if two solutions have the same NDR (i.e., they both belong to the same front), then in order to maintain the diversity, the solution located in a lesser crowded region in the front is preferred. If the crowding distance is the same for two solutions, then any of the solutions is assigned a higher ranking, randomly. The crowding distance measure is used as a tiebreaker in this selection technique, called the crowded tournament selection operator. In more simpler terms, if the solutions are in the same non-dominated front, the solution with a higher crowding distance is the winner.

Procedure 3: Crowding distance assignment
Step 1 Set the front counter $k = 1$
Step 2 For each solution X_i in the set P_k, first assign $cd_i = 0$
Step 3 For each objective function f_m, $m = 1, 2, \ldots M$, sort the set P_k in ascending order of its objective function value
Step 4 Assign $cd_1^m = cd_L^m = \infty$, where $L = |P_k|$
Step 5 For all other solutions $X_j \in P_k$, $j = 2, 3, \ldots, L - 1$, assign crowding distance using Eq. (3.11)
Step 6 Calculate the final crowding distance value (CD_i) for each solution ($X_i, i = 1 \ldots N$) using Eq. (3.12).

The pseudo-code of the discussed MOSCA algorithm is shown in Algorithm 1.

Algorithm 1 Multi-objective sine cosine algorithm [17]

Generate P_o randomly in S and evaluate f for the generated P_o
Sort the P_o based on the elitist non-dominated sorting method and find the NDR and fronts
Compute CD for each front
Update solutions ($X_j \in P_o$) using SCA algorithm to generate new population P'
Merge P_o and P' to create a new population P_{new}
For P_{new} perform step 2
Based on NDR and CD sort P_{new}
Replace P_o with P_{new} for the first N members of P_{new}.

3.3.2.2 Multi-objective Sine Cosine Algorithm for Optimal DG Allocation Problem

Raut and Mishra [19] developed another Pareto-based multi-objective sine cosine algorithm (MOSCA) to address the issues of optimal distribution generators (DGs) allocation. This approach applies a fast non-dominated sorting approach and the crowding distance operator. In addition to this, to enhance the performance, r_1 of SCA is defined as an exponential decreasing parameter and a self-adapting levy mutation as defined in Eqs. (3.13) and (3.14) is adopted.

$$r_1 = b \times e^{(-t/T)} \tag{3.13}$$

$$P_{g,j}^{t+1} = P_{g,j}^{t} + \text{levy} \times A(j) P_{g,j}^{t} \tag{3.14}$$

where $P_{g,j}^{t}$ is the value of the best agent in the jth dimension, levy step length is calculated from Eq. (3.15), and a self-adapting control coefficient A is calculated using Eqs. (3.18), (3.19) and (3.20).

$$\text{levy} = 0.01 \times \left(\frac{S \times \sigma}{T^{(\frac{1}{\alpha})}} \right) \tag{3.15}$$

where S and T are random numbers in the range [0, 1]. σ is defined as:

$$\sigma = \left(\frac{\Gamma(1+\alpha) \times \sin(\frac{\pi\alpha}{2})}{\Gamma(\frac{1+\alpha}{2}) \times \alpha \times 2^{(\frac{\alpha-1}{2})}} \right)^{\frac{1}{\alpha}} \tag{3.16}$$

where

$$\Gamma(k) = (k-1)! \tag{3.17}$$

The large value of A in the early iteration enhances the exploration, while the gradual decrease in A with increasing iteration numbers facilitates the exploitation.

$$A(j) = e^{(\frac{-\epsilon \times t}{T})(1 - \frac{w(j)}{w_{\max}(j)})} \tag{3.18}$$

$$w(j) = \left| P_{\text{best},j}^{t} - \left(\frac{1}{N} \sum_{i=1}^{N} X_{i,j}^{t} \right) \right| \tag{3.19}$$

$$w_{\max}(j) = \max(P_j^t) - \min(P_j^t) \tag{3.20}$$

where ϵ and α are constants, $w(j)$ is the difference between the jth dimension value of the current best solution and jth dimension average value of the population. $w_{\max}(j)$ is the maximum distance of the best solution from the worst solution.

Once the Pareto optimal set of non-dominated solutions is obtained, a fuzzy-based mechanism is employed to extract the best trade-off solutions from the obtained Pareto set and assist the decision-making process. Due to the imprecise nature of the decision-maker's judgment, each objective function is represented by a membership function. A simple linear membership function μ_l^k is defined for each objective and the membership value of kth solution in jth objective is given as

$$\mu_l^k = \frac{F_l^{\max} - F_l^k}{F_l^{\max} - F_l^{\min}} \tag{3.21}$$

where μ is the fuzzy membership function, $F_l^{\min}$ and $F_l^{\min}$ are the maximum and minimum values of lth objective function. For each member of non-dominated set, the normalized membership value (μ^k) is calculated using the following equation:

$$\mu^k = \frac{\sum_{l=1}^{m} \mu_l^k}{\sum_{k=1}^{K} \sum_{l=1}^{m} \mu_l^k} \qquad (3.22)$$

where K is the total number of Pareto solutions. The maximum value of μ^k is selected as the Pareto optimal solution.

3.3.2.3 Multi-objective Sine Cosine Algorithm for Spatial-Spectral Clustering Problem

Wan et al. [20] developed a multi-objective SCA for remote sensing image spatial-spectral clustering (MOSCA_SSC) that uses a knee-point-based selection approach [32], the concept of Pareto dominance and elitism. 'Knees' are the solutions of the Pareto front in which any modification to improve one objective would significantly deteriorate at least one other objective. The technique of Pareto dominance combined with elitism ensures that the non-dominated solutions survive in the succeeding generations of the algorithm. A multi-objective model consisting of multiple clustering objectives is utilized for the purpose of the clustering task of remote sensing image data. In MOSCA_SSC, two widely used metrics for remote sensing data, namely the Xie-Beni (XB) index and Jeffries–Matusita (Jm) distance combined with the spatial information are used as objective functions for the optimization purposes [33] (see Eqs. 3.23 and 3.24)

$$(XB)_{ind} = \frac{\sum_{i=1}^{K} \sum_{j=1}^{N} \mu_{ij}^m ||x_j - U_i||^2}{N \min_{i \neq k} ||U_i - U_j||^2} \qquad (3.23)$$

$$(SJm)_{ind} = \sum_{i=1}^{K} \sum_{j=1}^{N} \mu_{ij}^m ||x_j - U_i||^2 + \phi \sum_{i=1}^{K} \sum_{j=1}^{N} \mu_{ij}^m ||\overline{x_j} - U_i||^2 \qquad (3.24)$$

where K is the number of cluster centers, N is the total number of pixels in the remote sensing image, m is the fuzzy weighting exponent, which determines the degree of sharing of samples between classes. x_j is a vector, which denotes the jth pixel of the image, and μ_{ij} denotes the fuzzy membership. U_i and U_k, are the jth and the kth cluster centers, and ϕ is the control parameter. $\overline{x_j}$ represents the average gray value [33].

The procedure for MOSCA_SSC is described as follows:

Main Steps of the MOSCA_SSC

Step 1 Initialize a set of parent search agents (population) of size NP.

Step 2 Select the initial destination point using the Fuzzy C-Means (FCM) method.

Step 3 Generate new offsprings using SCA to get a new population and merge the new population with the old population to get $2 \times$ NP solutions.

Step 4 For each search agent, calculate the values of the two clustering objective functions using Eqs. (3.23) and (3.24).

Step 5 Rank the parent and the offspring search agents using the non-dominance sorting and crowding distance approach and select the NP best solutions from $2 \times$ NP solutions.

Step 6 Select the destination point using the knee-point-based selection approach.

Step 7 Repeat the process from steps 3 to 6 until the stopping criteria is reached.

The Fuzzy C-Means (FCM) method [34], mentioned in the step 2 of the MOSCA_SSC is used to obtain the initial destination point, as SCA requires initial destination points to begin the optimization procedure. Knee-point selection approach is utilized for automatically updating the destination points in the SCA algorithm [32]. In non-dominance diversity-based approaches, there are two challenging aspects to handle. The first aspect of this approach is to produce multiple non-dominated solutions to form a near optimal Pareto front, while the second aspect is to maintain diversity among these non-dominated solutions. Researchers have proposed different methods and techniques to tackle this challenge. The use of an external archive to store the non-dominated solutions and a grid mechanism to improve the diversity of the non-dominated solutions are some major methods to enhance the capabilities of non-dominance diversity-based approaches.

Archive: Archive is a storage memory where non-dominated solutions of previous iterations are stored. The non-dominated solutions stored in archive can be utilized for further generating new solutions, and based on the dominance status of these newly generated solutions, the solutions stored in the archive are managed.

Grid Mechanism: It concerns with managing the diversity in the non-dominated solutions by locating the crowded region where non-dominated solutions lie. Different grid mechanism techniques are available in the literature for this purpose. However, the basic idea behind the grid mechanism is to divide the objective space into smaller regions or grids to observe the distribution of the non-dominated solutions. If the distance between non-dominated solutions is small, and the number of non-dominated solutions is big, that particular grid is considered crowded.

Selim et al. [21] proposed a multi-objective sine cosine algorithm with an external archive and adaptive grid mechanism to handle the DSTATCOM allocation problem as mentioned below.

3.3.2.4 Multi-objective Sine Cosine Algorithm for DSTATCOM Problem

In distribution systems, to improve the voltage profile, and overall reliability, Distribution STATic COMpensators (DSTATCOMs) are used. Selim et al. [21] proposed multi-objective SCA (MOSCA) and used fuzzy logic decision-making to optimally install multiple Distribution STATic COMpensators (DSTATCOMs). The optimization procedure is carried out to determine the optimum size and location of DSTAT-COMs that leads to the minimization of power loss, voltage deviation (VD), and maximization of the voltage stability index (VSI) of the radial distribution system. MOSCA is a Pareto-based algorithm that utilizes the Pareto ranking scheme in the sine cosine algorithm to handle this multi-objective optimization problem. MOSCA incorporates an external archive of solutions to keep the historical record of non-dominated solutions, and the mechanism of adaptive grid [12] to maintain the diversity of the non-dominated solutions in the external archive. The major objective of the external archive and grid mechanism is the fact that a solution that is non-dominated with respect to its current population might not be non-dominated with respect to other solutions stored in the archive from the previous iterations in the evolutionary process. In MOSCA [21], an archive controller and adaptive grid mechanism are employed to store the non-dominated solutions and maintain the diversity of the solutions.

Archive Controller
Archives controllers are responsible for deciding whether a solution should be included in the archive or not. The non-dominated solutions generated at each iteration of the MOSCA are compared with the solutions inside the archive [21]. This archive is initially empty, and with the iterations, non-dominated solutions are added and updated. However, a fixed size of the archive is maintained because of memory limitations. If the archive is empty, then the candidate solution is accepted. If the archive is not empty, there are three possibilities—if solutions in the archive dominate the new candidate solution, it is not added in the archive. If there are solutions in the archive that are dominated by the new solution, then those solutions are eliminated. If the new candidate solution is neither dominated by any solution of the archive nor dominates any solution, it is added to the archive depending on the availability of the slot in the archive. Finally, the adaptive grid mechanism is triggered if the external population has exceeded its permitted capacity [12]. The archiving behavior is summarized in Algorithm 2. The graphical illustration of the archive update mechanism is depicted in Fig. 3.6.

Adaptive Grid Mechanism
The adaptive grid mechanism maintains the diversity in non-dominated solutions lying in the archive. It is used to delete solutions from the external archive if the external population has reached its maximum size. The MOSCA proposed in [21] utilizes the adaptive grid mechanism proposed in [12] to generate well-distributed Pareto fronts. This mechanism measures the degree of crowding in different regions of the solution space. The objective space region is divided into $d \times M$ number of

Algorithm 2 Pseudo-code for maintaining an archive of Pareto solutions

for Each candidate solution X_i in the new population **do**
   if A candidate dominates any solutions in the archive **then**
      if Archive is not full **then**
         Remove all dominated members from the archive and add the candidate to the archive
      else if the archive is full **then**
         if the candidate increases the diversity in the archive **then**
            Run the grid mechanism to remove the solution with the highest frequency from the
            archive and add the candidate search agent to the archive;
         end if
      end if
   else
      Reject the candidate
   end if
end for

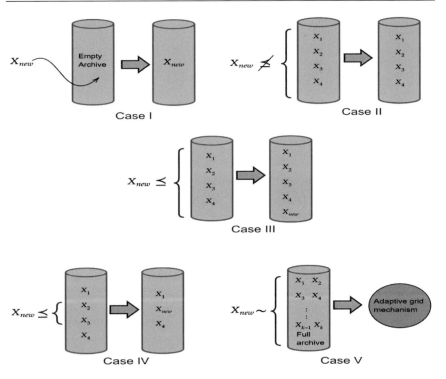

Fig. 3.6 Archive update mechanism

equal-sized M-dimensional hyper-cubes, where d is a user-defined parameter that
denotes the number of grids (see Fig. 3.7). The archived solutions are placed in these
hyper-cubes according to their locations in the objective space. A map of the grid is
maintained, to calculate the number of non-dominated solutions lying in a particular
grid. If the archive is already full, then the new solution cannot be included without

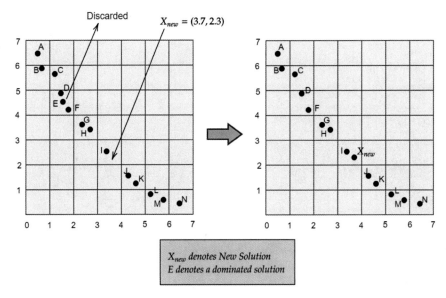

Fig. 3.7 Graphical representation of the insertion of a new element in the adaptive grid when the individual lies within the current boundaries of the grid [12]

making the space in the archive. In this case, the hypercube with the highest number of solutions is identified, and if the new solution does not belong to this hypercube, it is included in the archive, and one of the solutions from the archive is eliminated. If the new solution is not dominated or dominates any other solution in the archive, while the archive is full, the solution with the highest grid count is deleted from the archive. If the new solution inserted into the archive lies outside the current bounds of the grid, then the grid is recalculated, and each solution inside it is relocated (see Fig. 3.8).

Algorithm 3 Pseudo-code for adaptive grid mechanism

Search Space S, Number of grids d, Archive A, Grid M_i $i = 1, 2 \ldots d$
A grid counter n_c denoting the number of solutions in a particular grid M_i, Initialize $n_c = 0$
for Each non-dominated solution X_i in the archive A **do**
  if Solution $X_i \in M_i$ **then**
    Increase the grid counter by 1
    Calculate the number of solutions in each grid
  else if the archive is full **then**
    if a new solution is eligible for entering the archive **then**
      Remove a solution lying in a grid with a maximum n_c value
    end if
  else
    Recalculate the grid
  end if
end for

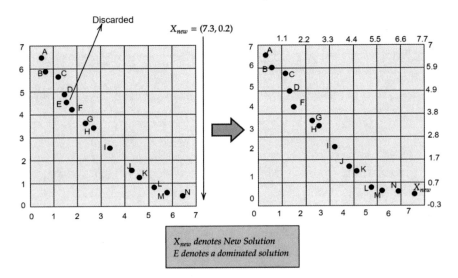

Fig. 3.8 Graphical representation of the insertion of a new element in the adaptive grid when it lies outside the previous boundaries of the grid [12]

When the non-dominated solution in the archive are assessed on the basis of crowding, the solutions with the least crowding or the solutions located in the least congested region of the objective space are given preference over the solutions lying in the more crowded region. The pseudo-code of the MOSCA [21] is given in Algorithm 4.

Algorithm 4 Pseudo-code of MOSCA [21]

Initialize the population $\{X_1, X_2, \ldots, X_N\}$ randomly
Initialize the parameter 'b'
For each candidate solution, calculate the objective function values
Find the non-dominated search agents and initialize the archive with them
Select the destination point from the archive
$t = 0$
while Termination criteria is met **do**
    for each search agent **do**
        Update the position of the search agent using the original SCA
    end for
    Calculate the objective values of all the search agents
    Find the non-dominated search agents
    Update the archive using Algorithm 2
    if any of the new added solutions to the archive is located outside the hypercube **then**
        Update the grids to cover the new solution(s)
    end if
    Select the destination point from the archive
    $t = t + 1$
end while
Return Archive

3.4   Conclusion

Optimization problems involving multiple objectives are common. In this context, meta-heuristics turn out to be a valuable tool, in particular, if the problem complexity prevents exact methods from being applicable and flexibility is required with respect to the problem formulation. Most real-world engineering problems involve simultaneously optimizing multi-objectives where considerations of trade-offs is important. Multi-objective sine cosine algorithm has shown its applicability to various application problems. Apart from basic MOO concepts, this chapter has covered various multi-objective sine cosine algorithms and their applications.

Practice Exercises

1. Prove that dominance relation is a partial order. (Hint: If a relation is reflexive, anti-symmetric, and transitive, it is called partial order.)
2. Given a set of points and a multi-objective optimization problem, analyze the statement that one point always dominates the others.
3. Given four points and their objective function values for multi-objective minimization:

$f_1(x_1) = 1, f_2(x_1) = 1, f_1(x_2) = 1, f_2(x_2) = 2, f_1(x_3) = 2, f_2(x_3) = 1,$
$f_1(x_4) = 2, f_2(x_4) = 2$

   (1) Which point dominates all the others?
   (2) Which point is non-dominated?
   (3) Which point is Pareto optimal?

4. Discuss the challenges involved in multi-objective optimization.
5. Comment on the dependence of the optimal solution on the weighting coefficients in the weighted sum approach.
6. For multi-objective optimization, the understanding of the Pareto front is very important. Explain.

References

1. X.-S. Yang, *Nature-Inspired Optimization Algorithms* (Academic Press, 2020)
2. K. Deb, Multi-objective optimisation using evolutionary algorithms: an introduction, in *Multi-Objective Evolutionary Optimisation for Product Design and Manufacturing* (Springer, 2011), pp. 3–34
3. M. Ehrgott, Vilfredo Pareto and multi-objective optimization. Doc. Math. 447–453 (2012)
4. M. Nagy, Y. Mansour, S. Abdelmohsen, Multi-objective optimization methods as a decision making strategy. Int. J. Eng. Res. Technol. (IJERT) **9**(3), 516–522 (2020)
5. K. Deb, *Multi-Objective Optimization Using Evolutionary Algorithms* (Wiley, 2014)
6. K. Miettinen, *Nonlinear Multiobjective Optimization*, vol. 12 (Springer Science & Business Media, 2012)

7. C.-L. Hwang, A.S.M. Masud, *Multiple Objective Decision Making—Methods and Applications: A State-of-the-Art Survey*, vol. 164 (Springer Science & Business Media, 2012)
8. R.S. Rosenberg, Stimulation of genetic populations with biochemical properties: I. The model. Math. Biosci. **7**(3–4), 223–257 (1970)
9. J. David Schaffer, Multiple objective optimization with vector evaluated genetic algorithms, in *Proceedings of the First International Conference of Genetic Algorithms and Their Application* (1985), pp. 93–100
10. N. Srinivas, K. Deb, Muiltiobjective optimization using nondominated sorting in genetic algorithms. Evol. Comput. **2**(3), 221–248 (1994)
11. K. Deb et al., A fast and elitist multiobjective genetic algorithm: NSGA-II. IEEE Trans. Evol. Comput. **6**(2), 182–197 (2002)
12. C.A.C. Coello, G.T. Pulido, M.S. Lechuga, Handling multiple objectives with particle swarm optimization. IEEE Trans. Evol. Comput. **8**(3), 256–279 (2004)
13. J.D. Knowles, D.W. Corne, Approximating the nondominated front using the Pareto archived evolution strategy. Evol. Comput. **8**(2), 149–172 (2000)
14. H.A. Abbass, R. Sarker, C. Newton, PDE: a Pareto-frontier differential evolution approach for multi-objective optimization problems, in *Proceedings of the 2001 Congress on Evolutionary Computation* (IEEE Cat. No. 01TH8546), vol. 2 (IEEE, 2001), pp. 971–978
15. L.A. Moncayo-Martinez, D.Z. Zhang, Multi-objective ant colony optimisation: a metaheuristic approach to supply chain design. Int. J. Prod. Econ. **131**(1), 407–420 (2011)
16. S. Mirjalili, Dragonfly algorithm: a new meta-heuristic optimization technique for solving single-objective, discrete, and multi-objective problems. Neural Comput. Appl. **27**(4), 1053–1073 (2016)
17. M.A. Tawhid, V. Savsani, Multi-objective sine cosine algorithm (MO-SCA) for multi-objective engineering design problems. Neural Comput. Appl. **31**(2), 915–929 (2019)
18. E. Zitzler, K. Deb, L. Thiele, Comparison of multiobjective evolutionary algorithms: empirical results. Evol. Comput. **8**(2), 173–195 (2000)
19. U. Raut, S. Mishra, A new Pareto multi-objective sine cosine algorithm for performance enhancement of radial distribution network by optimal allocation of distributed generators. Evol. Intell. **14**(4), 1635–1656 (2021)
20. Y. Wan et al., Multiobjective sine cosine algorithm for remote sensing image spatial-spectral clustering. IEEE Trans. Cybern. (2021)
21. A. Selim, S. Kamel, F. Jurado, Voltage profile enhancement using multi-objective sine cosine algorithm for optimal installation of DSTACOMs into distribution systems, in *2019 10th International Renewable Energy Congress (IREC)* (IEEE, 2019), pp. 1–6
22. A. Selim, S. Kamel, F. Jurado, Optimal allocation of distribution static compensators using a developed multi-objective sine cosine approach. Comput. Electr. Eng. **85**, 106671 (2020)
23. A. Selim, S. Kamel, F. Jurado, Efficient optimization technique for multiple DG allocation in distribution networks. Appl. Soft Comput. **86**, 105938 (2020)
24. D. Gonidakis, A. Vlachos, A new sine cosine algorithm for economic and emission dispatch problems with price penalty factors. J. Inf. Optim. Sci. **40**(3), 679–697 (2019)
25. P. Singh, R. Tiwari, Optimal allocation of STATCOM using improved sine cosine optimization algorithm, in *2018 8th IEEE India International Conference on Power Electronics (IICPE)* (IEEE, 2018), pp. 1–6
26. R.M. Rizk-Allah et al., A new sine cosine optimization algorithm for solving combined non-convex economic and emission power dispatch problems. Int. J. Energy Convers. **5**(6), 180–192 (2017)
27. R.M. Rizk-Allah, R.A. El-Sehiemy, A novel sine cosine approach for single and multiobjective emission/economic load dispatch problem, in *2018 International Conference on Innovative Trends in Computer Engineering (ITCE)* (IEEE, 2018), pp. 271–277
28. S. Mirjalili, SCA: a sine cosine algorithm for solving optimization problems. Knowl.-Based Syst. **96**, 120–133 (2016)
29. M. Sakawa, H. Yano, T. Yumine, An interactive fuzzy satisficing method for multiobjective linear-programming problems and its application. IEEE Trans. Syst. Man Cybern. **17**(4), 654–661 (1987)

30. L.A. Zadeh, G.J. Klir, B. Yuan, *Fuzzy Sets, Fuzzy Logic, and Fuzzy Systems: Selected Papers*, vol. 6 (World Scientific, 1996)
31. S. Krishnamurthy, R. Tzoneva, Impact of price penalty factors on the solution of the combined economic emission dispatch problem using cubic criterion functions, in *2012 IEEE Power and Energy Society General Meeting* (IEEE, 2012), pp. 1–9
32. X. Zhang, Y. Tian, Y. Jin, A knee point-driven evolutionary algorithm for many objective optimization. IEEE Trans. Evol. Comput. **19**(6), 761–776 (2014)
33. Y. Wan et al., Hyperspectral remote sensing image band selection via multi-objective sine cosine algorithm, in *IGARSS 2019–2019 IEEE International Geoscience and Remote Sensing Symposium* (IEEE, 2019), pp. 3796–3799
34. M.N. Ahmed et al., A modified fuzzy c-means algorithm for bias field estimation and segmentation of MRI data. IEEE Trans. Med. Imaging **21**(3), 193–199 (2002)

Chapter 4
Sine Cosine Algorithm for Discrete Optimization Problems

In many scenarios, the nature of the decision-making is discrete, and we have to deal with a situation where decisions have to be made from the set of discrete choices, or mutually exclusive alternatives. Choices like passing the electric signal versus not passing the electric signal, going upward versus downward, or choosing a certain route over other available routes are discrete in nature. There are many physical systems for which continuous variable modeling is not sufficient to handle the complexity of the physical systems. For instance, communication models, transportation models, finite element analysis, and network routing models are discrete models. The discrete nature of the search space offers the leverage of definiteness, and possibilities for graphical representation of given particular choices. In fact, discrete optimization problems are of paramount importance in various branches of sciences, like decision-making, information systems, and combinatorics. Operation management decision problems, like product distribution, manufacturing facility design, machine sequencing, and production scheduling problems, fall under the purview of discrete optimization problems. Network designing, circuit designing, and automated production systems are also represented as discrete optimization problems. Moreover, the application spectrum of discrete optimization problems includes data mining, data processing, cryptography, graph theory, and many others.

The decision space of the discrete optimization problems is either finite or similar to an enumerable set in the sense of cardinality. Mathematically, a general discrete optimization problem is given by Eq. (4.1),

$$\text{Min (or Max)} \quad C(X)$$
$$\text{subject to} \quad X \in P \tag{4.1}$$

where X is a permutation of decision variables, P is the set of feasible permutations or the search space, and $C(X)$ is the objective function. We are interested in finding an optimal permutation or arrangement $X \in P$, such that the objective value is optimal.

© The Author(s) 2023
J. C. Bansal et al., *Sine Cosine Algorithm for Optimization*,
SpringerBriefs in Computational Intelligence,
https://doi.org/10.1007/978-981-19-9722-8_4

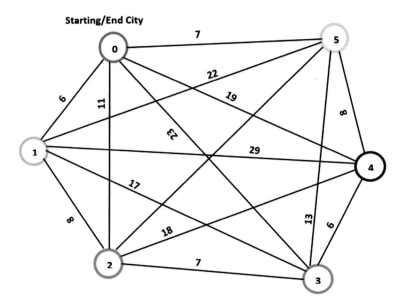

Fig. 4.1 Traveling salesman problem with 6 cities

The problem of finding optimal arrangement might look simple and easy to handle, but the discreteness of the problem brings forth a massive burden of dimensionality and computational complexity. For instance, traveling salesman problem (TSP) with 100 cities would require $(100 - 1)!$ permutations to be checked if the brute force method is applied. $99!$ would be approximately equal to 9.3326×10^{155}, and it is estimated that our observable universe has approximately 10^{82} atoms. But, to make our point clear, it involves a very huge computational cost, and of course, brute force method is not a recommended choice. The discrete optimization problems may look simple in formulation, but these problems can be computationally expensive. First, we explain a few discrete optimization models, then methods to solve them are included, and finally, discrete variants of SCA will be discussed in subsequent sections.

4.1   Discrete Optimization Models

We are mentioning some of the discrete optimization problems to give a brief idea to the reader. For more discrete optimization problems, a reader can refer to the book 'Discrete Optimization' by Parker and Rardin [1].

1. **Traveling Salesman Problem (TSP)**: Suppose a salesman wants to travel n number of cities lying in a geographical area. The salesman wants to start his journey from the present city and travel all the remaining $n - 1$ cities exactly

once and return back to his current location with minimizing the cost of the journey. This cost may include money, time, distance, or all. The problem can be visualized on a graph where each city represents a node (vertex). Arc (edge) lengths denote the associated cost of traveling between the cities. We have already discussed about the complexity of TSP with 100 cities. Mathematically, TSP can be formulated as:

For any directed (or, undirected) graph with certain fixed weights lying on the edges, we are interested in determining a closed cycle that includes each vertex of the graph exactly once, and this closed cycle would yield minimum total edge weight. Graphical illustration of TSP with 6 cities is given in Fig. 4.1.

The application spectrum of the TSP is very wide. A lot of well-known problems like vehicle routing, computer wiring, X-ray crystallography, crew-scheduling problem, and aircraft scheduling problems can be studied as the instance of a TSP problem [2].

In the literature, many exact and approximate methods are available for handling traveling salesman problem. One of the earliest approaches to solve TSP problem was proposed by Dantzig, Fulkerson, and Johnson (DFJ) [3]. DFJ algorithm formulates TSP into integer linear programming (ILP) problem with constraints, and prohibits the formation of subtours, i.e., tours containing less than n vertices [3]. Miller, Tucker, and Zemlin (MTZ) proposed an alternative formulation of ILP by reducing the number of subtour elimination constraints at the cost of introducing a new variable in the TSP problem [4]. Other than ILP approaches, branch-and-bound (BB) algorithms also proved effective in providing optimal solutions to the TSP problem. In BB algorithms, some of the problem constraints are relaxed initially, and at later stages, feasibility is regained by including constraints in an enumerative manner [5]. Many researchers including Eastman [6], Little et al. [7], Shapiro [8], Murty [9], Bellmore and Malone [10], Garfinkel [11], Smith et al. [12], Carpaneto and Toth [13], Balas and Christofides [14], and Miller and Pekny [15] proposed various branch-and-bound algorithms for handling TSP instances. However, high computational complexities of above mentioned approaches motivated researchers to employ heuristic and meta-heuristic approaches to solve TSP problems. For instance, ant colony optimizer (ACO) [16], particle swarm optimizer (PSO) [17], and discrete spider monkey optimizer (D-SMO) [18] are some popular heuristic approaches to produce optimal solution for TSP problems.

2. **Knapsack Problem**: In Knapsack problem, we are interested in finding a finite set $K \subseteq \mathcal{Z}$ of integer values k_i, where $i = 1, 2, \ldots n$ that minimizes $F(k) = f(k_1, k_2, \ldots, k_n)$ satisfying the restriction $g(k_1, k_2, \ldots, k_n) \geq v$, where v is a parameter.

The Knapsack problem is of particular interest in the various branches of sciences and decision-making problems like, resource allocation problems, portfolio allocation, capital budgeting, and project selection applications [19, 20]. The Knapsack problem has also been used in generating covering inequalities [21, 22], and in the area of cryptography [23]. Different versions of knapsack problems are available in the literature, for instance multi-dimensional knapsack problem (MKP), multiple choice knapsack problem (MCKP), and multi-dimensional multi-choice knapsack problem [24]. Traditional methods like dynamic programming, linear programming relaxation, Lagrangian relaxation, reduction methods, and branch-and-bound approaches are available in the literature to handle a variety of knapsack problems [25]. On the other hand, meta-heuristic approaches like simulated annealing (SA), genetic algorithm (GA), and particle swarm optimizer (PSO) have proved their capabilities in handling knapsack problems [26–28].

3. **Vertex Coloring**: Vertex coloring problem is a particular case of Vertex labeling problem in which vertices in a graph are labeled using colors. In this problem, the task is to label vertices of a given graph with a minimum number of colors, such that each vertex of the graph is in order that two adjacent vertices (an edge connecting vertices) are not labeled with the same color.

   Some major applications of the vertex coloring problem include scheduling tasks like job scheduling, aircraft scheduling, and time-table scheduling [29]. The assignment of radio frequencies, separating combustible chemical combinations, and handling multi-processor tasks are also instances of vertex coloring problems [30]. The traditional approaches like dynamic programming, branch-and-bound methods, and integer linear programs have been used in exact methods for handling the vertex coloring problems [31]. For instance, algorithms like Lawler's algorithm [32], Eppstein's algorithm [33], Byskov's algorithm [34], and Bodlaender and Kratsch algorithm [35] utilize dynamic programming approaches. And, Brelaz's algorithm [36], and Zykov's algorithm [37] are based on branch-and-bound methods. Meta-heuristic approaches like genetic algorithm (GA), simulated annealing (SA), ant colony optimizer (ACO), and cuckoo search (CS) have been utilized in the literature for solving vertex coloring problem [38, 39].

4. **Shortest Path Planning Problem**: The goal is to determine the shortest path connecting two fixed vertices in a given graph with certain fixed cost on the edges, such that the total length of a distinct sequence of edges that connects the two vertices is minimum.

   Shortest path planning problem has applicability in road networks, designing electric circuits, logistic communication, robotic path planning, etc. [40]. Dijkstra's algorithm [41], Floyd–Warshall algorithm [42], and Bellman–Ford algorithm [43] are some traditional algorithms in the literature of shortest path problem. Apart from the traditional approaches, genetic algorithms (GA), particle swarm optimizer (PSO), ant colony optimizer (ACO), and artificial bee colony (ABC) algorithms are popular meta-heuristic approaches to solve shortest path problem [44, 45].

5. **Set Covering**: The task is to find a family of subsets $\{P_i \subseteq P : i \in K\}$ (K is an index set) for a particular finite set P. These subsets P_i have a cost, say, c_i associated with them. One has to choose a collection of subsets such that the union of these subsets contains all the elements of the universal set P and the total cost of the collection is minimum.

   Set covering problem is a particular problem of interest for various disciplines, like operations research, computer science, and management. Crew scheduling problems, optimal location problems, optimal route selection problems are some of the instances of set covering problems [46]. The traditional approaches like linear programming relaxation, Lagrange relaxation, and branch-and-bound methods are available in the literature to tackle set covering problems [47, 48]. Meta-heuristic approaches including genetic algorithm (GA), ant colony optimizer (ACO), and XOR-based ABC have been utilized to solve the set covering problem [49, 50].

Discrete problems are widely used by different branches of sciences and industries. For instance, an airline company will be interested in solving TSP to optimize the route plan of their fleet. Similarly, the knapsack problem has a wide variety of applications in the financial modeling, production and inventory management, and optimal design for queueing networks model in manufacturing [51]. The detailed discussion of discrete optimization problems and their applications is beyond the scope of this book. The focus of this chapter is to present an overview of the discrete version of sine cosine algorithm (SCA) and its applications. But, before proceeding further, we briefly discuss about the discrete optimization methods.

4.2 Discrete Optimization Methods

Similar to continuous optimization methods, discrete optimization methods can also be studied under the two major categories of exact methods and approximate methods. The exact methods offer the guarantee of finding the optimal solution in a bounded time, but these methods are incapable of handling problems with large instances. Branch-and-cut, branch-and-bound, and Gomory's cutting plane method are some of the examples of exact methods. An interested reader can refer to the books 'Combinatorial Optimization' by Cook [52] and 'Discrete Optimization' by Parker and Rardin [1] to have a detailed idea about exact methods. On the other hand, approximate methods do not offer any guarantee for locating the optimal solution(s), but can produce near optimal solutions way much faster than the exact methods [53]. These heuristic methods are easy to implement and do not require extensive computational power to generate solutions. The greedy algorithms, sequential algorithms, heuristics local search, and methods based on random-cut, and randomized-backtracking are some of the examples of approximate algorithms. Figure 4.2 represents the classification of discrete optimization problems.

In the last three decades, researchers have tried to propose many efficient methods to tackle the discrete optimization problems. In the class of these efficient methods, the population-based meta-heuristic techniques have important role to play. Meta-heuristic possesses the potential to provide efficient near optimal solution(s) to these discrete optimization problem. Ant-colony optimizer [16], tabu search (TS) [54], simulated annealing [55], and genetic algorithm [56] are some of the meta-heuristic algorithms proposed for handling discrete optimization problems. The other class of meta-heuristic algorithms which were actually proposed for tackling continuous

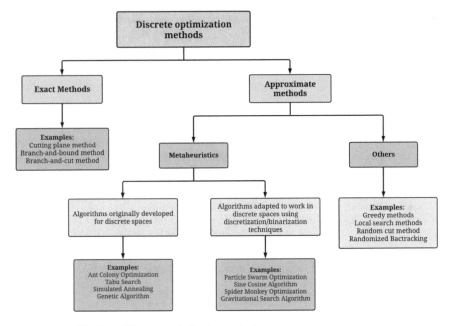

Fig. 4.2 Classification of discrete optimization methods

optimization problems, but later modified to solve the discrete optimization problems. Kennedy and Eberhart presented the discrete binary version of particle swarm optimization [57]. Discrete version of other meta-heuristic algorithms include discrete firefly-inspired algorithm [58], discrete teaching–learning-based optimization algorithm [59], binary coded firefly algorithm [60], binary magnetic optimization algorithm (BMOA) [61], binary cat swarm algorithm [62], binary dragonfly algorithm [63], and discrete spider monkey optimization [18].

Many discrete optimization problems can be reduced to binary optimization problems [1].

The sine cosine algorithm was originally proposed for continuous optimization problems [64]. The robust optimization capabilities of the SCA motivated the researchers for designing discrete sine cosine algorithm. In the next section, binarization techniques adopted to modify continuous sine cosine algorithm (SCA) will be discussed in detail.

4.3 Binary Versions of Sine Cosine Algorithm

In binary optimization problems, the decision variables can only take two values, typically 0 and 1. These two values can represent the True/False logic values, Yes/No, or On/Off. In general, the logical truth value is represented by '1' and false value is denoted by '0'. There are various techniques available to modify a continuous meta-heuristic into binary one. The binarization methods, like the nearest integer (NI) [65], the normalization technique [66], transfer functions [67], angle modulation [68], quantum approach [69], etc., are available in the literature to reinforce a binary version of continuous evolutionary or swarm intelligence algorithm [70]. In the literature of binarization techniques [53], two major categorization of binarization techniques were identified. The first category corresponds to a general techniques of binarization that enable the proceeding with the continuous meta-heuristics without altering the operators of the continuous algorithms. These techniques adopt mechanisms, like transfer functions [71], and angular modulation [68], to transform a continuous meta-heuristic algorithm into a binary version. Discrete PSO [17], binary coded firefly algorithm [60], binary magnetic optimization algorithm (BMOA) [61], and binary cat swarm algorithm [62] are some of the algorithms under the first category mentioned above. The second category consists of the techniques, in which the structure of meta-heuristics is altered. These methods rectifies the structure of the search space and hence reformulate the operators of the algorithms. Some techniques under this category include quantum binary algorithms [69], set-based approaches [72], techniques based on percentile concept [73], Db-scan unsupervised learning [74], and K-means transition ranking [75], to design binary versions of continuous meta-heuristic algorithms. Figure 4.3 depicts the various binarization techniques available in the literature.

4.3.1 Binary Sine Cosine Algorithm Using Round-Off Method

Hafez et al. [65] proposed a binary version of sine cosine algorithm (SCA) that utilizes the standard binarization rule. This proposed binary SCA was applied to feature selection problems. The goal is to choose combinations of features that maximize the classification performance and minimize the number of selected features. Therefore, the overall objective is to minimize the fitness value given by Eq. (4.2).

$$f_X = w * \epsilon + (1 - w)\frac{\sum_{i=1}^{n} x_i}{n} \qquad (4.2)$$

where f_X is the fitness function corresponding to a D-dimensional vector $X = (x_1, x_2, \ldots x_D)$, where $x_i = 0$ or 1. $x_i = 1$ represents the selection of the ith feature, while $x_i = 0$ indicate the non-selection of the ith feature. D is the total number of features in the given dataset. ϵ is the classifier error rate and w is a constant controlling the importance of classification performance to the number of features selected.

In the proposed approach, the range of all the decision variables is constrained to $\{0, 1\}$ using the rounding method, in which values of each decision variables are rounded to the integer value $0/1$ by employing Eq. (4.3). The features corresponding

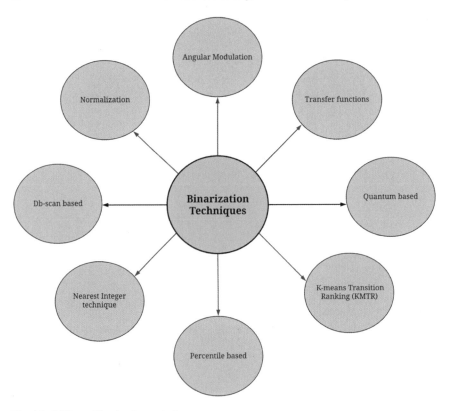

Fig. 4.3 Different binarization techniques

to the variable value 1 are selected. And, the features with variable value 0 are rejected.

$$X_{ij}^{t+1} = \begin{cases} 1 & \text{if } X_{ij}^t > 0.5 \\ 0 & \text{otherwise} \end{cases} \tag{4.3}$$

where X_{ij} is the value for ith search agent at the jth dimension.

The feature selection problem is of particular interest in machine learning algorithms. This is also an essential tool for attribute reduction or pre-processing of a large data sets. The least significant features which have very less relevance are removed to reduce the computational burden of the classification algorithm. It is quite evident that from the set of features we have to select or reject a feature. The choice available to us is of 'Yes/No' type, which fall under the purview of '0/1' discrete optimization problem or binary optimization. The proposed algorithm applied sine cosine algorithm (SCA) to find the combinations of features that have maximum impact on the classification performance.

4.3.2 Binary Sine Cosine Algorithm Using Transfer Functions

The procedure of converting a continuous meta-heuristic algorithm into its binary counterpart is called binarization. The technique of binarization using transfer functions has two major steps,

(1) The transfer function, which mapped the values generated by a continuous meta-heuristic algorithm to an interval $(0, 1)$.
(2) The binarization process, which consist of converting the real number lying in the interval $(0, 1)$ to a binary value.

Kennedy et al. [17] introduced transfer function (sigmoid function) for converting continuous PSO into discrete PSO. A transfer function facilitates the movement of ith search agent in the binary space by switching the value of jth dimension from 0 to 1 and vice versa. The advantage of utilizing a transfer function is that, it provides a probability of switching the solutions coordinates at a low computational cost [53]. In the literature of binarization techniques, there are several transfer functions available for converting a continuous meta-heuristic algorithm into its binary version [53]. The binary versions of a continuous meta-heuristic algorithms are constructed using the transfer functions, and these binary versions have the structure, similar to their continuous versions. The search agents' position is updated in the continuous space, and these continuous values are mapped in the interval $(0, 1)$ using transfer functions to generate a switching probability. These binary versions of continuous meta-heuristics differ from their continuous counterparts in the sense that, the search agents' position vector is a vector of binary digits rather than a vector of continuous values, and the position update mechanism is concerned with switching search agents' positions in the set $\{0, 1\}$ based on the transition probability obtained using the transfer function. The fundamental idea is to update the search agents' positions in such a way that the bit value of the search agents is switched between 0 and 1, with a probability based on the updated position of the search agents in the continuous space. The idea of using transfer function for converting a continuous meta-heuristic algorithm into a discrete meta-heuristic has also been incorporated in the sine cosine algorithm (SCA). The procedure for two-step binarization technique using transfer function method is graphically illustrated in the (Fig. 4.4).

Reddy et al. [71] proposed and investigated four binary variants of SCA to solve the binary natured profit-based unit commitment problem (PUCP). The proposed four variants use four different transfer functions for binary adaption of continuous search space and search agents. These transfer functions are mentioned as follows;

1. **The tangent hyperbolic transfer function (T)**

$$T(X^{t+1}) = \tanh(X^{t+1}) = \frac{e^{-(X^{t+1})} - 1}{e^{-(X^{t+1})} + 1} \qquad (4.4)$$

Mapping is given by:

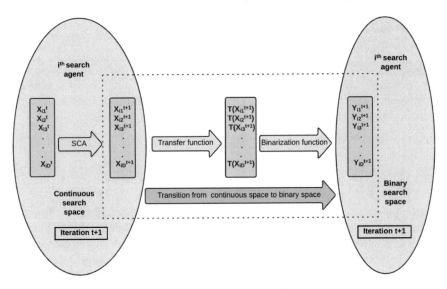

Fig. 4.4 Two-step binarization technique using transfer function

$$Y^{t+1} = \begin{cases} 0 & \text{if rand} < T(X^{t+1}) \\ 1 & \text{otherwise} \end{cases} \tag{4.5}$$

where X^t is the real-valued position of the search agent in the tth iteration and Y^t is its corresponding binary position. Here, rand is a uniformly distributed random number in the range $[0, 1]$.

2. **Sigmoidal transfer function (S)**

$$S(X^{t+1}) = \frac{1}{1 + e^{-X^{t+1}}} \tag{4.6}$$

Mapping is given by:

$$Y^{t+1} = \begin{cases} 0 & \text{if rand} < S(X^{t+1}) \\ 1 & \text{otherwise} \end{cases} \tag{4.7}$$

where rand has the same meaning as mentioned above.

3. **A modified sigmoidal transfer function (MS)**

$$MS(X^{t+1}) = \frac{1}{1 + e^{-10(X^{t+1}-0.5)}} \tag{4.8}$$

Mapping is given by:

$$Y^{t+1} = \begin{cases} 0 & \text{if rand} < S(X^{t+1}) \\ 1 & \text{otherwise} \end{cases} \tag{4.9}$$

where rand has the same meaning as mentioned above.

4. **Arctan transfer function (ArcT)**

$$\begin{aligned} \text{ArcT}(X^{t+1}) &= \arctan(X^{t+1}) \\ &= \left| \frac{2}{\pi} \arctan\left(\frac{\pi}{2} X^{t+1}\right) \right| \end{aligned} \tag{4.10}$$

Mapping is given by:

$$Y^{t+1} = \begin{cases} 0 & \text{if rand} < \text{ArcT}(X^{t+1}) \\ 1 & \text{otherwise} \end{cases} \tag{4.11}$$

where rand has the same meaning as mentioned above.

The performance of these different transfer functions to solve a binary profit-based unit commitment (PBUC) problem was investigated. The adequacy of the proposed approach, in terms of convergence and quality of solutions, was experimented over a benchmark test set. In terms of the solution quality, the arctan transfer function showed superior results out of all mentioned above variants, and the simple sigmoid transfer function could not produce satisfactory results.

Following the similar trend, Taghian et al. [67] proposed two other binary versions of SCA using the two-step binarization technique. The first version is called S-shaped binary sine cosine algorithm (SBSCA). In SBSCA, the S-shaped transfer function, defined in Eq. (4.12), is used to define a bounded probability of changing positions of the search agents [67].

$$S(X_{ij}^{t+1}) = \frac{1}{1 + e^{-X_{ij}^t}} \tag{4.12}$$

Then, the standard binarization rule, given in Eq. (4.13), is used to transform the solutions into a binary counterpart.

$$X_{ij}^{t+1} = \begin{cases} 1 & \text{if rand} < S(X_{ij}^{t+1}) \\ 0 & \text{otherwise} \end{cases} \tag{4.13}$$

Here, rand is a uniformly distributed random number in the range [0, 1]. The second version is called the V-shaped binary sine cosine algorithm (VBSCA) [67]. In VBSCA, the V-shaped transfer function is used to calculate the position changing probabilities given as:

$$V(X_{ij}^{t+1}) = \left| \frac{2}{\pi} \arctan\left(\frac{\pi}{2}\right)(X_{ij}^t) \right| \tag{4.14}$$

Then, the complement binarization rule, given by Eq. (4.15), is utilized to transform the solution into a binary domain.

$$X_{ij}^{t+1} = \begin{cases} \bar{X}_{ij}^{t+1} & \text{if rand} < V(X_{ij}^{t+1}) \\ (X_{ij}^t) & \text{otherwise} \end{cases} \qquad (4.15)$$

where $\bar{X}_{ij}^{t+1}$ represents the complement of X_{ij}^{t+1} at the iteration $t + 1$.

The performance of both the proposed algorithms was assessed and compared with four popular binary optimization algorithms, including binary GSA [76] and binary Bat algorithm [77], over five UCI medical datasets: pima, lymphography, heart, breast cancer, and breast-WDBC. The experimental results demonstrated that both the binary SCA variants have effectively enhanced the classification accuracy and yielded competitive or even better results when compared with the other existing algorithms.

4.3.3 Binary Sine Cosine Algorithm Using Percentile Concept

Another binary variant of SCA called binary percentile sine cosine algorithm (BPSCA) was introduced by Fernandez et al. [78], in which percentile concept was utilized to conduct the binary transformation of the sine cosine algorithm (SCA). In binary percentile concept, the magnitude of the displacement in jth component of a solution (say) X is calculated. Based on the magnitude of the displacement, these solutions are grouped in different percentile values of 20, 40, 60, 80, and 100. Solutions with least displacement values were grouped into 20-percentile value, while solutions with the maximum displacement were grouped into 100-percentile values [79].

The main issue using the binary percentile operator is that it may generate infeasible solutions in the search space. For handling the infeasible solutions, the BPSCA uses a heuristic operator. Heuristic operator chooses a new column, when solutions are needed to be repaired. As an input argument, the operator considers the set S_{in}, which is a set of columns to be repaired. The flowchart of the BPSCA algorithm is illustrated in Fig. 4.5.

To assess the performance of the percentile operator in obtaining the solutions, BPSCA was applied to solve the classic combinatorial problems called the set covering problem (SCP). The experimental results demonstrated that the percentile operator plays an important role in maintaining good-quality solutions. In addition, when BPSCA compared with the two best available meta-heuristic binary algorithms, namely jumping PSO (JPSO) [80] and multi-dynamic binary black hole (MDBBH) algorithm [81]. The experiments showed that the solutions obtained using BPSCA were similar to the jumping PSO. BPSCA generated superior results when compared with the MDBBH. The authors emphasized that, unlike JPSO, the percentile tech-

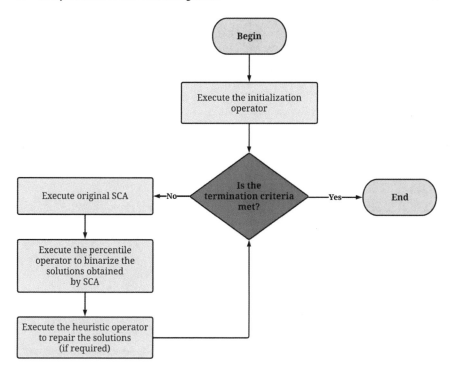

Fig. 4.5 Flowchart of the BPSCA

nique used in BPSCA allows binarization of any continuous meta-heuristic algorithm [78].

Pinto et al. [73] proposed percentile-based binary SCA (BPSCOA) using a repair operator instead of a heuristic operator. The repair operator handles the infeasible solutions produced during the optimization process. For repairing a particular solution, the coordinate with the maximum displacement measure is selected and eliminated from the solution [73]. The process is continued till the feasible solution is obtained. After this, repaired solution is improved by incorporating new elements in the solution such that no constraints are violated [73]. The flowchart of the BPSCOA is given in Fig. 4.6. In binarization process, the utility of percentile concept was evaluated by applying it to resolve the multi-dimensional knapsack problem (MKP). The results showed that the operator improved the precision and the quality of the solutions. The proposed method was contrasted with the binary artificial algae (BAAA) [82] and K-means transition ranking (KMTR) algorithms [75].

Till now, all the discrete optimization problems discussed above have binary nature. The solution(s) to these problems were 'Yes/No', or '0/1' type. Binary optimization problems hold important position in the discrete world of choices. However, sometimes we are interested in finding integer solutions to the real-world discrete optimization problems. Now, instead of making Boolean choices, we try to find the

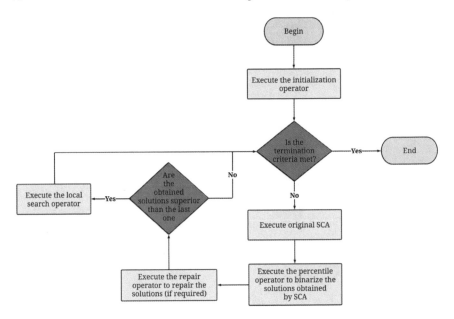

Fig. 4.6 Flowchart of the BPSCOA

solution(s) having integer values associated with the discrete optimization problem. In the next section, we will be discussing a general discrete version of sine cosine algorithm, in which solutions can take any finite integer value including the case of '0/1' type binary optimization problems.

4.4 Discrete Versions of Sine Cosine Algorithm

Tawhid et al. [83] proposed discrete version of sine cosine algorithm (DSCA) for solving traveling salesman problem (TSP). The objective function of the TSP is given by:

$$\text{Min } F(C) = \sum_{i=1, j=1}^{n} C_{i,j} \tag{4.16}$$

where the cost $C_{i,j}$ represents the Euclidean distance between any two towns i and j. For solving TSP, the authors adopted two local search techniques—the heuristic crossover [84] and the 2-opt [85] method on the best solution based on two randomly generated numbers between 0 and 1 (say R_1 and R_2), in a manner mentioned below,

If $R_1 < R_2$, Heuristic crossover is operated

Otherwise, 2-opt local search is operated

The psuedo-code of DSCA is given by Algorithm 1.

Algorithm 1 Pseudo-code of DSCA (Tawhid et al. [83])

Initialize the population of Hamiltonian cycles $\{C_1, C_2, \ldots, C_N\}$, randomly
Initialize the parameters associated with SCA
Calculate the objective function value for each search agent in the population
Identify the best solution obtained so far as the destination point C_b
initialize $t = 0$, where t is the iteration counter
while termination criteria is not met **do**
    Calculate r_1, using Eq. (2.4) and generate the parameters r_2, r_3, r_4, randomly
    for each search agent (C_i^t) **do**
        Generate the new search agents (C_i^{t+1}) using Eq. (2.7)
        if C_i^{t+1} is better than C_i^t **then**
            Replace C_i^t with C_i^{t+1}
        else
            Reject C_i^{t+1} and keep the C_i^t
        end if
    end for
    Update the current best solution (or destination point) C_b
    Generate random number R_1 and R_2
    if $R_1 < R_2$ **then**
        Update C_b using heuristic crossover
    else
        Update C_b using 2-opt local search
    end if
    $t = t + 1$
end while
Return the best solution C_b

The DSCA was tested on 41 different benchmark instances of symmetrical TSP. The results indicated that the technique provided optimal solutions for 27 benchmark problems and near optimal solutions for the remaining ones. When the results were compared with other state-of-the-art techniques, the DSCA demonstrated promising and competitive performance over the others.

Gholizadeh et al. [86] proposed discrete version of the sine cosine algorithm to tackle the discrete truss structures optimization problem and called it discrete modified SCA. In this algorithm, the solutions obtained from the traditional SCA are rounded to their nearest integer to speed up the process of optimization.

$$X_{ij}^{t+1} = \begin{cases} \text{round}(X_{ij}^t + r_1 \times \sin(r_2) \times |r_3 \times P_{gj}^t - X_{ij}^t|) & \text{if } r_4 < 0.5 \\ \\ \text{round}(X_{ij}^t + r_1 \times \cos(r_2) \times |r_3 \times P_{gj}^t - X_{ij}^t|) & \text{if } r_4 \geq 0.5 \end{cases} \qquad (4.17)$$

where round(·) truncate the values to their nearest integer. However, the solutions generated by the unintelligent round-offs might lie in an infeasible region, and their fitness values might differ drastically from that of the optimal solutions. The two main strategies—regeneration and mutation operator—are incorporated to address the issue of infeasible solutions. These two strategies help the algorithm to explore and exploit the design space in more robust manner. In the regeneration strategy, individual solutions of the population of size N are first sorted on the basis of objective function values in ascending order as follows:

$$\text{sort}(X^t) = [X_1^t, X_2^t \dots X_{k-1}^t, X_k^t, X_{k+2}^t \dots X_{N-1}^t, X_N^t] \tag{4.18}$$

where sort(X^t) is the current sorted population, and X_k^t to X_N^t are the worst solutions at iteration (t), that are required to be regenerated. Then, $\lambda \times N$ number of worst search agents (X_k^t to X_N^t) are removed from the population, where λ is a user defined parameter whose value lies in the interval (0, 1). The best solution found so far $X^* = [X_1^*, X_2^* \dots X_j^*, \dots, X_D^*]$ are then copied $\lambda \times N$ times in the population, that is,

$$\text{For } l = k, k+1 \dots N, \text{ Replace } X_l^t \text{ by } X_l^t = [X_1^*, X_2^* \dots X_D^*] \tag{4.19}$$

A randomly selected dimension $j \in [1, 2 \dots D]$ of each solution X_k^t to X_{N-1}^t is regenerated in a random manner using Eq. (4.20),

$$X_j^* = \text{round}\left(X_j^L + r \times (X_j^U - X_j^L)\right) \tag{4.20}$$

where X_j^L and X_j^U are lower and upper bounds of the jth dimension, and r is a random number in [0, 1].

The regenerated variables of the individuals X_k^t to X_{N-1}^t are then substituted in the last particle (X_N^t) to increase the chances of finding promising regions in the search space. In the second strategy, a mutation operator is applied in the generated solutions to escape the local optimal region. For each particle ($X_i, i = 1, 2 \dots N$), a random number in [0, 1] is generated, and if for the ith particle, the selected random number is less than a pre-defined mutation rate (mr), X_i will be regenerated using the following equation:

$$X_i^{t+1} = \text{round}\left(X_i^t + \left(\frac{t}{T}\right) \times R^t \otimes (X_{\text{best}}^t - X_r^t)\right) \tag{4.21}$$

where $\otimes$ denotes the vector product and R^t is a D dimensional vector of random numbers in the range [0, 1] at the tth iteration. X_{best}^t is the best solution of the current population, and X_r^t is a randomly selected solution from the current population at iteration t. The values of λ and mr were taken to be 0.2 and 0.05, respectively. The pseudo-code of the algorithm is given in Algorithm 2.

The proposed algorithm was applied on five well-known benchmark truss optimization problems, and the outcomes were compared with the original SCA and

---

Algorithm 2 Pseudo-code of discrete modified SCA

Initialize the population of search agents X_i $i = 1, 2, \ldots N$, randomly
Initialize the parameters associated with SCA
Calculate the objective function value for each search agent in the population
Identify the best solution obtained so far as the destination point P_g
initialize $t = 0$, where t is iteration counter
while termination criteria is not met **do**
    Calculate r_1, using Eq. (2.4) and generate the parameters r_2, r_3, r_4 randomly
    for each search agent (X_i^t) **do**
        Generate the new search agent (X_i^{t+1}) using Eq. (4.17)
        Identify and regenerate the worst search agents using Eqs. (4.18), (4.19) and (4.20)
        Mutate the search agents based on the mutation rate mr using Eq. (4.21)
    end for
    Update the current best solution (or destination point) P_g
    $t = t + 1$
end while
Return the best solution P_g

---

other optimization algorithms. Discrete modified SCA has shown superiority over other discrete algorithms [86].

Montoya et al. [87] proposed another discrete version of the sine cosine algorithm (SCA) to find the optimal location of the distributed generators (DGs) in alternating current (AC) distribution networks. An integer codification of SCA is proposed in the algorithm [88]. This codification technique eases the implementation with a matrix associated with the population. Any infeasible solutions obtained during the search process are removed from the population. In the technique of integer codification, each location of DGs is assigned a number between 2 and n (where n is the total number of nodes), and the number 1 is assigned to the slack node. The location of all the DGs can be visualized at the nodes, and an individual x_i does not appear more than once in any node to maintain the codification. The initial population for the proposed SCA will be represented in the form of a matrix with the dimensions NP $\times$ N, defined as follows:

$$P^t = \begin{bmatrix} x_1^t \\ x_2^t \\ \vdots \\ x_{NP}^t \end{bmatrix} \begin{bmatrix} x_{1,1} & x_{1,2} & \cdots & x_{1,N} \\ x_{2,1} & x_{2,2} & \cdots & x_{2,N} \\ \vdots & \vdots & \ddots & \vdots \\ x_{NP,1} & x_{NP,2} & \cdots & x_{NP,N} \end{bmatrix} \qquad (4.22)$$

where NP denotes the population size, and N is the number of DGs available for installation. In the initial population, ith search agent in the jth dimension $(x_{i,j})$ is randomly generated natural number between 2 and n, using the following equation:

$$x_{i,j} = \text{round}(2 + \text{rand} \times (n - 2)) \qquad (4.23)$$

where the round($\cdot$) denotes the floor function or nearest integer part of the given number, and rand is a normal distributed random number with mean 0 and standard deviation 1.

The following constraint must be satisfied to secure the feasible solution using the following equation:

$$x_{i,j} \neq x_{i,k}; \forall k = 1, 2, \ldots N \text{ and } k \neq j \qquad (4.24)$$

To preserve the feasible solution space at the time of the initializing population, the authors establish that every solution's component is different from one another. The discrete version of the sine cosine algorithm is summarized in Algorithm 3.

Algorithm 3 Pseudo-code of DSCA (Montoya et al. [87])

Initialize the population of search agents randomly using Eq. (4.23)
Initialize the parameters associated with SCA
Calculate the objective function value for each search agent in the population
Identify the best solution obtained so far as the destination point P_g
initialize $t = 0$, where t is iteration counter
while termination criteria is not met **do**
   Calculate r_1, using Eq. (2.4) and generate the parameters r_2, r_3, r_4 randomly
   for each search agent (x_i^t) **do**
      Update the position of the search agents using Eq. (2.7)
      for $j = 1 : N$ **do**
         if $(X_{i,j} < 2 \, || \, X_{i,j} > n)$ **then**
            Regenerate $X_{i,j}$ using Eq. (4.23)
         end if
      end for
   end for
   Calculate the objective function value for each new search agent
   Update the current best solution (or destination point) P_g
   $t = t + 1$
end while
Return the best solution P_g

Practice Exercises

1. How is the binary version of SCA different from the original SCA?
2. What are the major limitations of using the transfer function for discrete optimization problems?
3. Binary PSO has also used the transfer function. Compare and tell how the transfer function taken in binary PSO and the transfer function taken in SCA are different.
4. What is the difference between the sigmoid transformation function and the tan hyperbolic transformation function?

5. What are the important issues to be considered while designing a discrete version of a meta-heuristic algorithm?
6. Using the standard SCA and converting it into a binary or discrete value is easier than going through conversion and working in the discrete space. Explain why it cannot be adopted with the help of an example.

References

1. R.L. Rardin, R.G. Parker, *Discrete Optimization* (Academic Press, Inc., 1988)
2. D. Devendra, *Travelling Salesman Problem, Application and Theory*, vol. 1 (InTech, 2010)
3. G. Dantzig, R. Fulkerson, S. Johnson, Solution of the large-scale travelling salesman problem. Oper. Res. (1954)
4. C.E. Miller, A.W. Tucker, R.A. Zemlin, Integer programming formulation and travelling salesman problem. J. Assoc. Comput. Mach. (1960)
5. G. Laporte, The traveling salesman problem: an overview of exact and approximate algorithms. Eur. J. Oper. Res. (1992)
6. W.L. Eastman, Linear programming with pattern constraints, PhD thesis, Harvard University, Cambridge, 1958
7. J.D.C. Little, K.G. Murty, D.W. Sweeney, C. Karel, An algorithm for travelling salesman problem. Oper. Res. **11** (1963)
8. D.M. Shapiro, Algorithms for the solution of the optimal cost and bottleneck traveling salesman problems, Sc.D. thesis, Washington University, St. Louis, MO, 1966
9. K.G. Murty, An algorithm for ranking all the assignments in order of increasing cost. Oper. Res. **16** (1968)
10. M. Bellmore, J.C. Malone, Pathology of travelling-salesman subtour-elimination algorithms. Oper. Res. **19**, 278–307 (1971)
11. R.S. Garfinkel, On partitioning the feasible set in a branch-and-bound algorithm for the asymmetric traveling-salesman problem. Oper. Res. **21**, 340–343 (1973)
12. T.H.C. Smith, G.L. Thompson, V. Srinivasan, Computational performance of three subtour elimination algorithms for solving asymmetric traveling salesman problems. Ann. Discrete Math. **1**, 495–506 (1977)
13. G. Carpaneto, P. Toth, Some new branching and bounding criteria for the asymmetric travelling salesman problem. Manage. Sci. **26**, 736–743 (1980)
14. E. Balas, N. Christofides, A restricted Lagrangean approach to the traveling salesman problem. Math. Program. **21**, 19–46 (1981)
15. D.L. Miller, J.F. Pekny, Results from a parallel branch and bound algorithm for solving large asymmetric traveling salesman problems. Oper. Res. Lett. **8**, 129–135 (1989)
16. M. Dorigo, M. Birattari, C. Blum, M. Clerc, T. Stützle, A.F.T. Winfield, Ant colony optimization and swarm intelligence, in *5th International Workshop* (Springer, 2006)
17. J. Kennedy, R.C. Eberhart, A discrete binary version of the particle swarm algorithm, in *1997 IEEE International Conference on Systems, Man, and Cybernetics. Computational Cybernetics and Simulation*, vol. 5 (IEEE, 1997), pp. 4104–4108
18. M.A.H. Akhand, S.I. Ayon, S.A. Shahriyar, N.H. Siddique, H. Adeli, Discrete spider monkey optimization for travelling salesman problem. Appl. Soft Comput. J. **86**(4), 469–476 (2020)
19. J.H. Lorie, L.J. Savage, Three problems in capital rationing. J. Bus. **28**, 229–239 (1955)
20. R. Nauss, The zero-one knapsack problem with multiple-choice constraints. Eur. J. Oper. Res. **2**, 125–131 (1978)
21. E. Balas, E. Zemel, An algorithm for large zero-one knapsack problems. Oper. Res. **28**, 1130–1154 (1980)

22. L.A. Wolsey, Faces for a linear inequality in 0–1 variables. Math. Program. **8**, 165–178 (1975)
23. M. Merkle R. Hellman, Hiding information and signatures in trapdoor knapsacks. IEEE Trans. Inf. Theory **24**, 525–530 (1978)
24. C. Wilbaut, S. Hanafi, S. Salhi, A survey of effective heuristics and their application to a variety of knapsack problems. IMA J. Manag. Math. **19**, 227–244 (2008)
25. K. Dudziński, S. Walukiewicz, Exact methods for the knapsack problem and its generalizations. Eur. J. Oper. Res. **28**(1), 3–21 (1987)
26. A. Liu, J. Wang, G. Han, S. Wang, J. Wen, Improved simulated annealing algorithm solving for 0/1 knapsack problem, in *Sixth International Conference on Intelligent Systems Design and Applications, 2006. ISDA'06*, vol. 2 (IEEE, 2006)
27. F. Qian, R. Ding, Simulated annealing for the 0/1 multidimensional knapsack problem. Numer. Math. Engl. Ser. **16**(4), 320 (2007)
28. L. Ouyang, D. Wang, New particle swarm optimization algorithm for knapsack problem, in *8th International Conference on Natural Computation* (2012)
29. U. Ufuktepe, G.B. Turan, Applications of graph coloring, in *Lecture Notes in Computer Science* (2005)
30. P. Gupta, O. Sikhwal, A study of vertex—edge coloring techniques with application. Int. J. Core Eng. Manag. (IJCEM) **1**(2) (2014)
31. A.M. de Lima, R. Carmo, Exact algorithms for the graph coloring problem. Rev. Inform. Teór. Apl. (RITA) **25** (2018). ISSN 2175-2745
32. E. Lawler, A note on the complexity of the chromatic number problem. Inf. Process. Lett. **5**(3), 66–67 (1976)
33. D. Eppstein, Small maximal independent sets and faster exact graph coloring. J. Graph Algorithms Appl. **7**(2), 131–140 (2003)
34. J.M. Byskov, Chromatic number in time O(2.4023n) using maximal independent sets. BRICS Rep. Ser. **9**(45), 1–9 (2002)
35. H.L. Bodlaender, D. Kratsch, An exact algorithm for graph coloring with polynomial memory. UU-CS, vol. 2006, no. 15, pp. 1–5 (2006)
36. D. Brelaz, New methods to color the vertices of a graph. Commun. Appl. Comput. Mach. **22**(4), 251–256 (1979)
37. A. Zykov, On some properties of linear complexes. Mat. Sb. (N.S.) **24**(66)(2), 418–419 (1962)
38. A. Layeb, H. Djelloul, S. Chikhi, Quantum inspired cuckoo search algorithm for graph colouring problem. Int. J. Bio-Inspired Comput. **7**, 183–194 (2015)
39. A. Kole, D. De, A.J. Pal, Solving graph coloring problem using ant colony optimization, simulated annealing and quantum annealing—a comparative study, in *Studies in Computational Intelligence*, vol. 1029 (Springer, 2022)
40. M. Kairanbay, H.M. Jani, A review and evaluations of shortest path algorithms. Int. J. Sci. Technol. Res. **2**(6) (2013)
41. E.W. Dijkstra, A note on two problems in connexion with graphs. Numer. Math. 269–271 (1959)
42. R.W. Floyd, Algorithm 97 shortest path. Commun. ACM **5**, 345 (1962)
43. R. Bellman, On a routing problem. Q. J. Appl. Math. **16**, 87–90 (1958)
44. D.D. Caprio, A. Ebrahimnejad, H. Alrezaamiri, F. Santos-Arteaga, A novel ant colony algorithm for solving shortest path problems with fuzzy arc weights. Alex. Eng. J. **61**(5) (2022)
45. M. Gen, R. Cheng, D. Wang, Genetic algorithms for solving shortest path problems, in *Proceedings of 1997 IEEE International Conference on Evolutionary Computation (ICEC '97)* (1997)
46. A. Caprara, P. Toth, M.A. Fischetti, Algorithms for the set covering problem. Ann. Oper. Res. **98**, 353–371 (2000)
47. E. Balas, A class of location, distribution and scheduling problems: modelling and solutions methods, in *Proceedings of the Chinese-US Symposium on System Analysis* (Wiley, 1983)
48. E. Balas, M.C. Carrera, A dynamic subgradient-based branch-and-bound procedure for set covering. Oper. Res. **44**, 875–890 (1996)

49. R. Soto et al., A XOR-based ABC algorithm for solving set covering problems, in *The 1st International Conference on Advanced Intelligent System and Informatics (AISI2015)*, Beni Suef, Egypt, 28–30 Nov 2015 (Springer, 2016), pp. 209–218
50. K.S. Al-Sultan, M.F. Hussain, J. Nizami, A genetic algorithm for the set covering problem. J. Oper. Res. Soc. **47**, 702–709 (1996)
51. K.M. Bretthauer, B. Shetty, The nonlinear knapsack problem—algorithms and applications. Eur. J. Oper. Res. **1**(1), 1–14 (2002)
52. W.J. Cook, W.H. Cunningham, *Combinatorial Optimization* (Wiley, 1998)
53. B. Crawford et al., Putting continuous metaheuristics to work in binary search spaces. Complexity **2017** (2017)
54. F. Glover, Future paths for integer programming and links to artificial intelligence. Comput. Oper. Res. **13**(5), 533–549 (1986)
55. S. Kirkpatrick, C.D. Gelatt, Jr., M.P. Vecchi, Optimization by simulated annealing. Science **220**(4598), 671–680 (1983)
56. M. Mitchell, *An Introduction to Genetic Algorithms* (MIT Press, 1998)
57. J. Kennedy, R.C. Eberhart, A discrete binary version of the particle swarm algorithm, in *1997 IEEE Conference on Systems, Man, and Cybernetics* (1997)
58. M.K. Sayadi, A. Hafezalkotob, S.G.J. Naini, Firefly-inspired algorithm for discrete optimization problems: an application to manufacturing cell formation. J. Manuf. Syst. **32**(1), 78–84 (2013)
59. A. Lotfipour, H. Afrakhte, A discrete teaching-learning-based optimization algorithm to solve distribution system reconfiguration in presence of distributed generation. Int. J. Electr. Power Energy Syst. **82**, 264–273 (2016)
60. B. Crawford et al., A binary coded firefly algorithm that solves the set covering problem. Roman. J. Inf. Sci. Technol. **17**(3), 252–264 (2014)
61. S.A. Mirjalili, S.Z.M. Hashim, BMOA: binary magnetic optimization algorithm. Int. J. Mach. Learn. Comput. **2**(3), 204 (2012)
62. B. Crawford et al., Binary cat swarm optimization for the set covering problem, in *2015 10th Iberian Conference on Information Systems and Technologies (CISTI)* (IEEE, 2015), pp. 1–4
63. M. Mafarja et al., Binary dragonfly optimization for feature selection using time-varying transfer functions. Knowl.-Based Syst. **161**, 185–204 (2018)
64. S. Mirjalili, SCA: a sine cosine algorithm for solving optimization problems. Knowl.-Based Syst. **96**, 120–133 (2016)
65. A.I. Hafez et al., Sine cosine optimization algorithm for feature selection, in *2016 International Symposium on Innovations in Intelligent Systems and Applications (INISTA)* (IEEE, 2016), pp. 1–5
66. A.P. Engelbrecht, G. Pampara, Binary differential evolution strategies, in *2007 IEEE Congress on Evolutionary Computation* (IEEE, 2007), pp. 1942–1947
67. S. Taghian, M.H. Nadimi-Shahraki, Binary sine cosine algorithms for feature selection from medical data. arXiv preprint arXiv:1911.07805 (2019)
68. B.J. Leonard, A.P. Engelbrecht, C.W. Cleghorn, Critical considerations on angle modulated particle swarm optimisers. Swarm Intell. **9**(4), 291–314 (2015)
69. J. Sun, B. Feng, W. Xu, Particle swarm optimization with particles having quantum behavior, in *Proceedings of the 2004 Congress on Evolutionary Computation* (IEEE Cat. No. 04TH8753), vol. 1 (IEEE, 2004), pp. 325–331
70. Z.A. El Moiz Dahi, C. Mezioud, A. Draa, Binary bat algorithm: on the efficiency of mapping functions when handling binary problems using continuous-variable-based metaheuristics, in *IFIP International Conference on Computer Science and Its Applications* (Springer, 2015), pp. 3–14
71. K.S. Reddy et al., A new binary variant of sine cosine algorithm: development and application to solve profit-based unit commitment problem. Arab. J. Sci. Eng. **43**(8), pp. 4041–4056 (2018)
72. Y.-J. Gong et al., Optimizing the vehicle routing problem with time windows: a discrete particle swarm optimization approach. IEEE Trans. Syst. Man Cybern. Part C (Appl. Rev.) **42**(2), 254–267 (2011)

73. H. Pinto et al., A binary sine cosine algorithm applied to the knapsack problem, in *Computer Science On-line Conference* (Springer, 2019), pp. 128–138
74. J. García et al., A Db-scan binarization algorithm applied to matrix covering problems. Comput. Intell. Neurosci. **2019** (2019)
75. J. García et al., A k-means binarization framework applied to multidimensional knapsack problem. Appl. Intell. **48**(2), 357–380 (2018)
76. E. Rashedi, H. Nezamabadi-Pour, S. Saryazdi, GSA: a gravitational search algorithm. Inf. Sci. **179**(13), 2232–2248 (2009)
77. S. Mirjalili, S.M. Mirjalili, X.-S. Yang, Binary bat algorithm. Neural Comput. Appl. **25**(3), 663–681 (2014)
78. A. Fernéndez et al., A binary percentile sin cosine optimisation algorithm applied to the set covering problem, in *Proceedings of the Computational Methods in Systems and Software* (Springer, 2018), pp. 285–295
79. J. García et al., A percentile transition ranking algorithm applied to binarization of continuous swarm intelligence metaheuristics, in *International Conference on Soft Computing and Data Mining* (Springer, 2018), pp. 3–13
80. S. Balaji, N. Revathi, A new approach for solving set covering problem using jumping particle swarm optimization method. Nat. Comput. **15**(3), 503–517 (2016)
81. J. García et al., A multi dynamic binary black hole algorithm applied to set covering problem, in *International Conference on Harmony Search Algorithm* (Springer, 2017), pp. 42–51
82. X. Zhang et al., Binary artificial algae algorithm for multidimensional knapsack problems. Appl. Soft Comput. **43**, 583–595 (2016)
83. M.A. Tawhid, P. Savsani, Discrete sine cosine algorithm (DSCA) with local search for solving traveling salesman problem. Arab. J. Sci. Eng. **44**(4), 3669–3679 (2019)
84. W.-P. Liu et al., Hybrid crossover operator based on pattern, in *2011 Seventh International Conference on Natural Computation*, vol. 2 (IEEE, 2011), pp. 1097–1100
85. G.A. Croes, A method for solving traveling-salesman problems. Oper. Res. **6**(6), 791–812 (1958)
86. S. Gholizadeh, R. Sojoudizadeh, Modified sine cosine algorithm for sizing optimization of truss structures with discrete design variables. Iran Univ. Sci. Technol. **9**(2), 195–212 (2019)
87. O.D. Montoya et al. A hybrid approach based on SOCP and the discrete version of the SCA for optimal placement and sizing DGs in AC distribution networks. Electronics **10**(1), 26 (2020)
88. O.D. Montoya, W. Gil-González, C. Orozco-Henao, Vortex search and Chu-Beasley genetic algorithms for optimal location and sizing of distributed generators in distribution networks: a novel hybrid approach. Eng. Sci. Technol. Int. J. **23**(6), 1351–1363 (2020)

Chapter 5
Advancements in the Sine Cosine Algorithm

In the last few decades, the development and advancement of meta-heuristic algorithms have become the focus of the research community as these algorithms face various challenges like, balance between exploration and exploitation, tuning of parameters, getting trapped in local optima, and very slow convergence rate. Sine Cosine Algorithm (SCA) also faces similar kinds of challenges and sometimes fails to perform effectively in finding the global optimal solution. Sine and Cosine are trigonometric operators with a 90° phase shift from each other. The range of sine and cosine functions lies in the range $[-1, 1]$. Sine and cosine functions in the position update equation of SCA help solutions to perform search procedure. However, in some situations, SCA promotes similar solutions in the search space, which results in the loss of diversity in the population, and the search process is susceptible to trapping in the region of local optimum [1]. Motivated by these challenges, SCA has been modified to improve its capability and efficiency in several ways. Several strategies have been employed to alter the basic version of SCA [2], aiming to enhance its effectiveness and optimization capabilities. In this chapter, we will discuss about these modifications and strategies, which have been incorporated into the sine cosine algorithm (SCA) in past few years. Apart from this, we will briefly describe the applications of the modified versions of SCA.

The modifications and ensemble of new strategies into the SCA algorithm include—modification in the update mechanism, change in the parameters involved, the introduction of elitism, the introduction of new operators, the introduction of an encoding scheme, the introduction of several statistical distributions for random number generations, etc. For the sake of brevity, we will briefly describe about these modifications and developments in the following manner,

1. Modifications in the position update mechanism
2. Opposition-based learning (OBL) in SCA
3. Quantum-inspired SCA
4. Hybridization of SCA with other meta-heuristics.

© The Author(s) 2023
J. C. Bansal et al., *Sine Cosine Algorithm for Optimization*,
SpringerBriefs in Computational Intelligence,
https://doi.org/10.1007/978-981-19-9722-8_5

5.1 Modifications in the Position Update Mechanism

The position update mechanism or position update operator can be considered as the core of any population-based meta-heuristic algorithm. The movement of the search agents in the search space is controlled by the position update mechanism. It is responsible for updating the current position of the search agents in an intelligent stochastic manner. In the literature of SCA, various modifications in the position update mechanism have been proposed to modify SCA in different ways.

Long et al. [1] proposed an improved version of the SCA (ISCA) for solving high-dimensional problems. This approach is inspired by the integration of the inertia weight (w) in the particle swarm optimizer (PSO) [3]. In this approach, the position update equation is modified by including the concept of inertia weight coefficient (w) to speed up the convergence and prevent local optima entrapment. Furthermore, a new nonlinearly decreasing conversion parameter based on the Gaussian function is introduced to keep the fine-tune balance between SCA's exploration and exploitation phases. The suggested modifications in the position update equation is given by Eq. (5.1).

$$
X_{ij}^{t+1} = \begin{cases} w(t) \cdot X_{ij}^t + r_1 \cdot \sin(r_2) \times \left| r_3 \cdot P_{gj}^t - X_{ij}^t \right| & \text{if } r_4 < 0.5 \\ w(t) \cdot X_{ij}^t + r_1 \cdot \cos(r_2) \times \left| r_3 \cdot P_{gj}^t - X_{ij}^t \right| & \text{if } r_4 \geq 0.5 \end{cases} \tag{5.1}
$$

where $w \in [0, 1]$ is the inertia weight coefficient. The value of w is linearly decreased from the initial value (w_s) to the final value (w_e) according to the following equation:

$$
w(t + 1) = w_e + (w_s - w_e) \times \frac{(T - t)}{t} \tag{5.2}
$$

where T denotes the maximum number of iterations, and t is the current iteration number.

Along with the introduction of the weight coefficient, Long et al. [1] proposed modifications in the control parameter r_1. The control parameter r_1 is the critical control parameter in the SCA algorithm which helps in controlling the exploration and exploitation phase of the algorithm by controlling the step size. The linearly decreasing value of r_1 helps the algorithm in choosing large step sizes in the initial phase and small step sizes at later phases of the optimization process [2]. However, the linearly decreasing value of r_1 might restrict its convergence rate and accuracy. Long et al. [1] presented a new nonlinearly decreasing strategy for control parameter r_1 based on the Gaussian function, mentioned in Eq. (5.3).

$$
r_1(t) = a_e + (a_s - a_e) \times \exp\left(\frac{-t^2}{(m \times T)^2} \right) \tag{5.3}
$$

where t indicates the current iteration, T indicates the maximum number of iterations, m is the nonlinear modulation index, and a_s and a_e are the initial and final values of constant a, respectively.

Suid et al. [4] proposed modifications in its update position and in the control parameter r_1 by utilizing the mean of the best search agent's position and the position of the current search agent. In this approach, each agent updates its position dimension-wise with respect to the average of its current position and the best search agent's position to avoid premature convergence. The modified position update equation is given in Eq. (5.4).

$$X_{ij}^{t+1} = \begin{cases} \frac{X_{ij}^t + P_{gj}^t}{2} + r_1 \cdot \sin(r_2) \times \left| r_3 \cdot P_{gj}^t - X_{ij}^t \right| & \text{if } r_4 < 0.5 \\ \frac{X_{ij}^t + P_{gj}^t}{2} + r_1 \cdot \cos(r_2) \times \left| r_3 \cdot P_{gj}^t - X_{ij}^t \right| & \text{if } r_4 \geq 0.5 \end{cases} \tag{5.4}$$

The control parameter r_1 is updated using a nonlinear decreasing mechanism, instead of the linearly decreasing mechanism, as mentioned in Eq. (5.5).

$$r_1 = b \cdot \left(1 - \left(\frac{t}{T} \right)^{\alpha} \right)^{\beta} \tag{5.5}$$

where b is the constant parameter $(a = 2)$, T denotes the number of maximum iteration, t is the current iteration, and both α and β are positive real numbers.

Kumar et al. [5] proposed Weibull Pareto sine cosine optimization algorithm (WPSCO), a modification in the sine cosine algorithm (SCA) to solve the peak power detection problem in solar PV panels. In WPSCO, Weibull and Pareto distributions functions are integrated with the SCA algorithm in the position update equation, which improves the convergence rate and enhances the exploitation of the search spaces [5]. In the first stage, the SCA is applied to find the optimal place for all variables (see Eq. 5.6).

$$\Phi = \begin{cases} X_{ij}^t + r_1 \cdot \sin(r_2) \times \left| r_3 \cdot P_{gj}^t - X_{ij}^t \right| & \text{if } r_4 < 0.5 \\ X_{ij}^t + r_1 \cdot \cos(r_2) \times \left| r_3 \cdot P_{gj}^t - X_{ij}^t \right| & \text{if } r_4 \geq 0.5 \end{cases} \tag{5.6}$$

In the second stage, the positions of all variables are analyzed by the Weibull and Pareto distribution function, and the worst regions of the search space are filtered. The position update mechanism for the second stage is given in Eq. (5.7).

$$X_{ij}^{t+1} = \Phi \times \left[1 + \Omega \times \left\{ \frac{t}{T} \times W(1, \varepsilon) + \left(1 - \frac{t}{T} \right) \times P(0, \varepsilon, 0) \right\} \right] \tag{5.7}$$

where Ω is the inertia constant. $W(1, \varepsilon)$ is Weibull random number. $P(0, \varepsilon, 0)$ is Pareto random number. ε is the error modulation index described as:

$$\varepsilon = \frac{\left| P_{gj}^{t} - P_{gj}^{t-1} \right|}{\rho_{\max j}^{t}} \qquad (5.8)$$

where

$$\rho_{\max j}^{t} = \begin{cases} 1 & \text{if } t = 1 \\ \left| P_{gj}^{t} - P_{gj}^{t-1} \right| & \text{if } \rho_{\max}^{t-1} < \left| P_{gj}^{t} - P_{gj}^{t-1} \right| & \text{if } t = 1 \\ \rho_{\max j}^{t-1} & \text{otherwise} \end{cases} \qquad (5.9)$$

For maximum power point tracking (MPPT) of partially shaded PV system, Kumar et al. [6] proposed another variant of SCA, called Cauchy and Gaussian sine cosine optimization (CGSCO) algorithm. The CGSCO algorithm combines the Cauchy density [7] and Gaussian distribution function (GCF) [8] with the sine cosine algorithm (SCA). In the proposed method, firstly initial population is updated using the position update mechanism of the SCA algorithm, and then Cauchy and Gaussian mutation mechanisms are employed on the updated population matrix Φ at every iteration. The Cauchy-Gauss mutation mechanism is given in Eq. (5.10).

$$X_{\text{new}} = \Phi + [1 + \delta \times \{\eta \times N(0, 1) + (1 - \eta) \times C(0, 1)\}] \qquad (5.10)$$

where $N(0, 1)$ and $C(0, 1)$ are Gaussian and Cauchy random numbers, δ is an inertia constant, $\eta = \frac{t}{T}$, t is the current iteration, and T is the maximum number of iterations. The Cauchy density function enhances the global exploration ability and prevents the algorithm from trapping into the region of local minima. And, the Gaussian distribution function increases the local exploitation capabilities to enhance the rate of convergence of the proposed CGSCO algorithm.

In the position update mechanism, random components are drawn from different distributions. For example, normal distribution, Gaussian distribution, or Cauchy distribution play a very important role in managing the stochasticity of the underlying meta-heuristic algorithm. These random components are responsible for the movement in the search agent's position in the search space by deciding direction and step lengths randomly. In simpler terms, position update mechanisms can be considered as random walks followed by the agents or particles in the search space. A Lévy flight is a specific class of random walk in which the step lengths have a heavy-tailed probability distribution, that is, agents will take a large step sizes occasionally, which in turn improves the exploration capabilities of the underlying algorithm and helps the search agents in escaping local optimal regions of the search space [9].

Inspired by the concept of Lévy flight, Attia et al. [10] proposed a modified sine cosine technique for solving the optimal power flow (OPF) problem by embedding Lévy flight into the position update mechanism of the sine cosine algorithm (SCA). The introduction of Lévy flights in the position update mechanism enhances the global search capabilities of the algorithm, and prevents the agents from being trapped in the regions of local optima. In addition, a fine-tuning capability (i.e., adaptive tuning of population size strategy) is utilized, in which the size of the population is updated in the following manner:

if

$f_{\min}(t) < \{f_{\min}(t - 1), f_{\min}(t - 2), f_{\min}(t - 3), f_{\min}(t - 4)\}$, here t is iteration counter.

Then,

population size = Number of search agents in the $(t - 1)$th iteration $\times (1 - \alpha)$

else

population size does not change

α is a constant whose value is taken to be 0.05. This adaptive strategy for the population size provides a fast convergence rate to the proposed algorithm.

Similarly, inspired by the concept of Lévy flights, Qu et al. [11] proposed another SCA variant involving Lévy flight. For maintaining a better balance between the exploration and exploitation capabilities of the algorithm, the method of exponentially decreasing conversion (see Eq. 5.11) was applied to the control parameter r_1, and the method of linearly decreasing inertia weight (see Eq. 5.12) was adopted on w. This helps in achieving a smooth transition from global exploration to local development.

$$r_1 = b \cdot e^{\frac{t}{T}} \tag{5.11}$$

$$w = w_{\max} - (w_{\max} - w_{\min}) \cdot \frac{t}{T} \tag{5.12}$$

Here, T is the maximum number of iterations, t is the current iteration. $w_{\max}$ and $w_{\min}$ denote the maximum and minimum value of the weight parameter, respectively.

Along with the adaptive control parameter strategy, a random neighborhood search strategy is employed, in which a random solution in the vicinity of the optimal solution is used in the position update equation. This allows the algorithm to quickly jump out of the local optimum and increases the diversity of the population. The modified position update equation is mentioned in Eq. (5.13).

$$X_{ij}^{t+1} = \begin{cases} w \cdot X_{ij}^t + r_1 \cdot \sin(r_2) \times | r_3 \cdot P_{gj}^t \times (1 + \lambda \cdot \text{rand}(-1, 1)) - X_{ij}^t | & \text{if } r_4 < 0.5 \\ w \cdot X_{ij}^t + r_1 \cdot \cos(r_2) \times | r_3 \cdot P_{gj}^t \times (1 + \lambda \cdot \text{rand}(-1, 1)) - X_{ij}^t | & \text{if } r_4 \geq 0.5 \end{cases} \tag{5.13}$$

where r_1 and w are the same as mentioned in Eqs. (5.12) and (5.13), respectively. λ is a constant parameter, and r_2, r_3, and r_4 are control parameters.

A self-adapting greedy Lévy mutation strategy is applied to perturb the optimal solution to enhance the local exploitation ability of the algorithm, and to eliminate the defect of low efficiency in a later period [11]. The optimal solution is updated using the following equation:

$$P_{gj}^{t+1} = P_{gj}^t + \eta(j) \cdot L \cdot P_{gj}^t \tag{5.14}$$

Here, L is a random number drawn from Lévy distribution. P_{gj} is the optimal solution, g is a solution index, and j denotes the dimension at iteration counter t. $\eta(j)$ is the coefficient of self-adapting variation defined in Eq. (5.15).

$$\eta(j) = e^{\left(-\varepsilon \cdot \frac{t}{T}\right)\left(1 - \frac{r(j)}{r_{\max}(j)}\right)} \tag{5.15}$$

Here in Eq. (5.15), ε is a control parameter whose value is chosen to be 30. $r(j)$ denotes the adjusted optimal solution's position given by Eq. (5.16), and $r_{\max}$ denotes the difference between the maximum and minimum value of all the solutions in dimension j (see Eq. 5.17)

$$r(j) = P_{gj}^t - \frac{1}{N} \cdot \sum_{i=1}^{N} X_{i,j}^t \tag{5.16}$$

$$r_{\max} = \max\left(X_{.,j}^t\right) - \min\left(X_{.,j}^t\right) \tag{5.17}$$

where N is the population size, t is the current iteration, and T denotes the maximum iteration.

5.2 Opposition-Based Learning Inspired Sine Cosine Algorithm

In this section, we briefly discuss about the concept of opposition-based learning in the sine cosine algorithm. Opposition-based learning (OBL) is a search strategy proposed by Tizhoosh [12] for machine learning applications. It takes into consideration the opposite position of solutions in the search space to increase the chance of finding better solutions in the search space. For a given population, say X, the opposition-based population $\overline{X}$ is calculated in a given manner. Suppose $X_i = [x_{i,1}, x_{i,2}, \ldots, x_{i,D}]$ is a solution in X, then $\overline{X}_i$ is calculated using the following equation:

$$\overline{x_{ij}} = u_j + l_j - x_{ij}, \quad i = 1, 2 \ldots N; \quad j = 1, 2 \ldots D \tag{5.18}$$

where u_j and l_j are the upper and lower bounds of jth dimension, respectively.

The concept of OBL increases the chances of better exploration in the search space and utilizing the opposite positions of solutions in the search space helps in generating a more refined population. For instance, suppose X is a randomly generated population of size N. Using the concept of OBL, a new population $\overline{X}$ is generated using X. Now, there are $2N$ solutions in the search space, and out of these $2N$ solutions, N solutions are selected on the basis of fitness value. That is, fitness values of X and $\overline{X}$ are calculated, and N number of solutions with better fitness values is retained in the population, and the rest of the solutions are eliminated or deleted. For the sake of brevity, two modifications of SCA algorithm using the concept of opposition-based learning (OBL) are discussed below.

Elaziz et al. [13] proposed opposition-based sine cosine algorithm (OBSCA). The authors combined the opposition-based learning strategy with SCA in both the initialization phase and updating phase. In the initialization phase, a randomly generated population (say, X) containing N solutions is initialized, and the concept of OBL is employed to generate the opposition-based population (say, $\overline{X}$). The fitness values of both X and $\overline{X}$ are calculated and N better solutions are retained for the updating phase. In the updating phase, the population is updated using the SCA algorithm, and the opposition-based learning is employed in the updated population. The fitness values of both the population are calculated, and N better solutions are retained for the next iterations, and the rest of the solutions are eliminated. The iterative process is repeated until the termination criteria is satisfied.

Chen et al. [14] proposed a multi-strategy enhanced sine cosine algorithm based on Nelder–Mead simplex (NMs) [15] concept and the opposition-based learning (OBL) strategy for the parameter estimation of photovoltaic models. The Nelder–Mead simplex method is used to deal with unconstrained minimization problems and nonlinear optimization problems. It is a derivative-free direct search method based on functional value comparison. In every iteration, the algorithm first executes the SCA algorithm for updating the population, and then, the OBL mechanism is employed to diversify the population in order to enhance the exploration capability of the algorithm. After the OBL method, the NMs mechanism is incorporated as a local search technique on every solution in order to exploit the potential neighborhood regions of the search space. In detail, the best solution found after using the OBL mechanism in the current population is selected to construct a primary simplex. Then, the simplex is updated according to the NMs simplex mechanism for some k number of iterations, and then the algorithm switched back to the SCA algorithm. The k is a vital parameter whose value is chosen to be $D + 1$, if the optimization problem is D dimensional. The concept of OBL enhances the diversity of the population and benefits the exploration capabilities of the meta-heuristic algorithms. For more applications in the field of soft computing, machine learning, and fuzzy systems, an interested reader can refer to the literature review of opposition-based learning by Mahdavi et al. [16].

5.3 Quantum-Inspired Sine Cosine Algorithm

Apart from the above-mentioned strategies and techniques, researchers have also employed many other methods to modify the sine cosine algorithm. The quantum-inspired meta-heuristics are also becoming popular in recent times. Quantum-inspired meta-heuristics take their inspiration from the various quantum mechanics principles like superposition, uncertainty, inference, entanglement, etc., to model various optimization algorithms [17–19]. The concept of quantum computing, like quantum bits (Q-bits), quantum gates (Q-gates), and their superposition have been combined with various existing meta-heuristic algorithms like particle swarm optimizer [17], gravitational search algorithm [18], gray wolf optimizer [19], to incorporate the merits of quantum computing to some extent. Inspired by the concept of quantum computing, Fu et al. proposed chaos quantum sine cosine algorithm (CQSCA) [20]. In the proposed algorithm, a chaotic initialization method and the quantum concept of superposition are used to improve the performance of the SCA algorithm. CQSCA is employed to produce the optimal values for the parameters involved in the support vector machine (SVM) in order to recognize the pattern of different kinds of faults. In the proposed algorithm, the population is initialized with a chaotic variable using a duffing system to enhance the quality of searching global optima [21]. The dynamical equation of the duffing system is given below:

$$x''(t) + \eta x'(t) - \xi x(t) + \mu x^3(t) = A \cdot \cos(\tau t) \tag{5.19}$$

where A is the amplitude of driving force. The coefficient η is the damping degree whose value is taken to be 0.1, and ξ is the toughness degree whose value is chosen as 1. μ is the nonlinearity of power and its value is taken to be 0.25. τ is the circular frequency of the driving force and its value is taken to be 2.

After the chaotic initialization, the inherent characteristics of the qubits and quantum gate concepts help in achieving a better balance between the exploration and exploitation phase of the search process. The proposed algorithm uses quantum bits or qubits[1] to encode the position of search agents in the search space to avoid premature convergence [20]. A qubit can be expressed by probability amplitude P_i using the following equation:

$$P_i = \begin{bmatrix} \cos(\theta) \\ \sin(\theta) \end{bmatrix} = \begin{bmatrix} p_i^c \\ p_i^s \end{bmatrix} \tag{5.20}$$

where θ denotes the phase shift of a qubit.

Every search agent occupies two positions in the search space, namely the sine position (p_i^s) and the cosine position (p_i^c), represented by Eqs. (5.21) and (5.22), respectively.

$$p_i^s = [\sin(\theta_{i1}), \sin(\theta_{i2}), \dots, \sin(\theta_{iD})] \tag{5.21}$$

$$p_i^c = [\cos(\theta_{i1}), \cos(\theta_{i2}), \dots, \cos(\theta_{iD})] \tag{5.22}$$

where $\theta_{ij} = 2\pi \times \alpha$, and α is a random number in the range [0, 1].

All the encoded search agents update their positions based on an update equation utilizing the features of the SCA algorithm and quantum mechanics. The movements in the search agents are implemented using quantum rotation gate. The position update mechanism of the proposed mechanism is given in Eq. (5.23).

$$P_{\text{inew}} = \begin{bmatrix} p_{\text{inew}}^c \\ p_{\text{inew}}^s \end{bmatrix} \tag{5.23}$$

---

[1] A qubit is the smallest unit of information in quantum theory.

where p_{inew}^c and p_{inew}^s are calculated using Eqs. (5.24) and (5.25).

$$p_{inew}^c = \left(\cos(\theta_{i1}^k + \Delta\theta_{i1}^{k+1}), \cos(\theta_{i2}^k + \Delta\theta_{i2}^{k+1}), \cdots \cos(\theta_{iD}^k + \Delta\theta_{iD}^{k+1})\right) \quad (5.24)$$

$$p_{inew}^s = \left(\sin(\theta_{i1}^k + \Delta\theta_{i1}^{k+1}), \sin(\theta_{i2}^k + \Delta\theta_{i2}^{k+1}), \cdots \sin(\theta_{iD}^k + \Delta\theta_{iD}^{k+1})\right) \quad (5.25)$$

$$\Delta\theta_{ij}^{k+1} = \begin{cases} r_1 \cdot \sin(r_2) \times \Delta\theta_g^k \\ r_1 \cdot \cos(r_2) \times \Delta\theta_g^k \end{cases} \quad (5.26)$$

and,

$$\Delta\theta_g = \begin{cases} 2\pi + \theta_{gj} - \theta_{ij}, & \theta_{gj} - \theta_{ij} < -\pi \\ \theta_{gj} - \theta_{ij}, & -\pi \le \theta_{gj} - \theta_{ij} < \pi \\ \theta_{gj} - \theta_{ij} - 2\pi, & \theta_{gj} - \theta_{ij} > \pi \end{cases} \quad (5.27)$$

Then a mutation operator with quantum non-gate is adopted to avoid local optimum and increase the population diversity [20]. For each search agent, a random number is generated between (0, 1) and is compared with the mutation probability p_m. Then the probability amplitudes of randomly chosen qubits are updated as follows:

$$P_i = \begin{cases} \begin{bmatrix} \cos(\theta_{ij}) \\ \sin(\theta_{ij}) \end{bmatrix} & \text{if } \text{rand}_i < p_m \\ \\ \begin{bmatrix} \sin(\theta_{ij}) \\ \cos(\theta_{ij}) \end{bmatrix} & \text{otherwise} \end{cases} \quad (5.28)$$

Similarly, Lv et al. [22] proposed a quantum encoding scheme inspired modification in the encoding scheme of the search agents in the SCA algorithm. In the proposed algorithm, instead of using real-valued coding for the search agents, the idea of quaternion coding is used. In quaternion encoding, a search agent is expressed as a hyper-complex number containing one real part and three imaginary parts. The real part and imaginary parts of a solution are updated in parallel using the position update mechanism of the SCA algorithm. Quaternions are super-complex numbers represented as mentioned in Eq. (5.29) [23]

$$q = a_0 + a_1 i + a_2 j + a_3 k \quad (5.29)$$

Here, a_0, a_1, a_2, and a_3 are real numbers, and i, j, and k are imaginary numbers following given algebraic rules (Eq. 5.30),

$$i \cdot j = k \qquad j \cdot i = -k \qquad j \cdot k = i$$
$$k \cdot j = -i \qquad k \cdot i = j \qquad i \cdot k = -j; \qquad (5.30)$$
$$i \cdot i = j \cdot j = k \cdot k = -1$$

The quaternion (q) described in Eq. (5.29) can be further simplified as,

$$q = a_0 + a_1 i + a_2 j + a_3 k = c + dj \qquad (5.31)$$

Here, both c and d are complex numbers, and j is an imaginary part defined as follows,

$$c = a_0 + a_1 i \qquad (5.32)$$

$$d = a_2 + a_3 i \qquad (5.33)$$

A random population of quaternion encoded search agents (Q) is initialized in the problem's definition domain $[L, U]$, using Eq. (5.34),

$$Q = Q_R + Q_I \cdot i \qquad (5.34)$$

where Q_R and Q_I are complex numbers given by the following equations (see Eqs. 5.35 and 5.36).

$$Q_R = Q_{RR} + Q_{RI} \cdot i = \rho_R \cos \theta_R + \rho_R \sin \theta_R \cdot i \qquad (5.35)$$

$$Q_I = Q_{IR} + Q_{II} \cdot i = \rho_I \cos \theta_I + \rho_I \sin \theta_I \cdot i \qquad (5.36)$$

where ρ_R, ρ_I are random numbers generated in the range,

$$\rho_R, \rho_I \in \left[0, \left(\frac{L-U}{2} \right) \right] \qquad (5.37)$$

and, θ_I, θ_R are random numbers generated in the range,

$$\theta_I, \theta_R \in [-2\pi, 2\pi] \qquad (5.38)$$

The quaternion encoded search agents (Q) could be converted to their real-value counterpart X using Eqs. (5.39)–(5.41),

$$X_R = \rho_R \mathrm{sgn} \left(\sin \left(\frac{Q_{RR}}{\rho_R} \right) \right) + \frac{L+U}{2} \qquad (5.39)$$

$$X_I = \rho_I \mathrm{sgn} \left(\sin \left(\frac{Q_{II}}{\rho_I} \right) \right) + \frac{L+U}{2} \qquad (5.40)$$

$$X = \sqrt{\left(X_R^2 + X_I^2\right)} \tag{5.41}$$

The position the search agents (Q_k) is updated using the following position update mechanism mentioned in the equation below:

$$Q_k^{t+1} = \begin{cases} Q_k^t + r_1 \times \sin(r_2) \times \left|r_3 \times QP_k^t - S_k^t\right| & \text{if } r_4 < 0.5 \\ Q_k^t + r_1 \times \sin(r_2) \times \left|r_3 \times QP_k^t - Q_k^t\right| & \text{if } r_4 \geq 0.5 \end{cases} \tag{5.42}$$

where $k = $ RR, IR, RI, II. And, QP represents the best solution obtained so far and is represented in the quaternion form as follows in Eq. (5.43),

$$QP = QP_R + QP_I \cdot i \tag{5.43}$$

Here, QP_R and QP_I are given as follows,

$$QP_R = QP_{RR} + QP_{RI} \cdot i \tag{5.44}$$

$$QP_I = QP_{IR} + QP_{II} \cdot i \tag{5.45}$$

The incorporation of quantum techniques in the sine cosine algorithm improves the exploration-exploitation capabilities of the algorithm. However, in digital computers, one can not exactly simulate the true nature of quantum computing. Despite of having the limitations, quantum-inspired techniques can be realized as effective techniques for advancements in the existing meta-heuristic algorithms.

5.4 Covariance Guided Sine Cosine Algorithm

Liu et al. [24] proposed an improved version of sine cosine algorithm called covariance guided sine cosine algorithm (COSCA). In COSCA, sine cosine algorithm (SCA) is embedded with the covariance concept to speed up its convergence, and the OBL mechanism to improve the diversity in the population. In every iteration, search agents in the population are sorted based on their fitness value in ascending order. The top $H = \lceil N/4 \rceil$ agents are selected to create a guiding population, say, (P_{Guide}). For updating the position of the agents, the position update mechanism of the SCA algorithm is utilized. After updating the position of every search agent in the population, the opposition-based learning (OBL) strategy is employed. The opposite positions of all the agents are calculated to form an opposite population. The best agents are selected from both the current population and its opposite population to proceed with the search process. The concept of covariance is utilized in the guided population P_{Guide}. The value of the covariance $C_{j,k}$ between any two dimensions j and k in the guided population is calculated using the following equation:

$$C_{j,k} = \frac{1}{H-1} \times \sum_{h=1}^{H} (G_{h,j} - \overline{G}_j) * (G_{h,k} - \overline{G}_k) \quad j, k = 1, 2, \ldots D \qquad (5.46)$$

where $G_{h,j}$, and $G_{h,k}$ represent the jth and kth dimensions of the hth variable in the guided population (P_{Guide}), respectively. $\overline{G}_j$ and $\overline{G}_k$ denote the mean of the jth and kth dimensions of the guided population. The value of the covariance $C_{j,k}$ forms a covariance matrix C with the size $D \times D$.

Further, eigenvalue decomposition is performed on the covariance matrix C as given in Eq. (5.47).

$$C = OM^2O^{\mathrm{T}} \qquad (5.47)$$

Here, M is a diagonal matrix whose elements are equals to the eigenvalues of C, O is an orthogonal matrix, such that each column of O comprises the orthogonal basis for each eigenvector of the covariance matrix C.

For ith search agent X_i, its candidate position Y_i is calculated using the following equation:

$$Y_i = \overline{G} + \sigma \cdot O \cdot M \cdot \gamma \qquad (5.48)$$

where $\overline{G}$ is the mean value of the guided population. σ is a zoom factor whose value is taken to be 1.5, γ is a D-dimensional random vector, and the range of each component of γ lies in the [0, 1]. If Y_i is better than X_i, X_i is replaced by Y_i, otherwise, X_i remains unchanged.

5.5   Hybridization of SCA with Other Meta-heuristics

In the context of meta-heuristic algorithms, hybridization refers to the process of integrating two or more existing meta-heuristic algorithms to form a new variant or hybrid algorithm. The hybrid algorithm produced by integrating two or more different algorithms are meant to report better performance when compared to the algorithms, which are used in the process of hybridization. The basic idea of merging two or more existing algorithms is to utilize the merits and strengths of used algorithms while improving their drawbacks. For instance, suppose an algorithm (say) $\mathcal{A}$ is known for better exploration capabilities but suffers from the drawback of weak exploitation, and on the other hand, a different algorithm (say) $\mathcal{B}$ owns better exploitation capabilities but is prone to get stuck in the region of local optimum. The hybrid algorithm (say) $\mathcal{C}$ produced by integrating algorithms $\mathcal{A}$ and $\mathcal{B}$ is supposed to contain the merits of both of the parent algorithms and perform relatively better when compared to both of the algorithms. However, utilizing the techniques of hybridization is a challenging task and requires careful analysis. The process of hybridization can also be achieved using different classical algorithms like simplex methods, Nelder-Mead simplex methods, and local random search techniques [25].

Sine cosine algorithm (SCA) holds the decent ability to achieve a fine balance between the exploration and exploitation phase. However, the performance of SCA can be enhanced using the technique of hybridization for any specific application-oriented problems. Sine cosine algorithm (SCA) has been successfully hybridized with the algorithms like particle swarm optimizer (PSO) [25, 26], genetic algorithm (GA) [27], differential evolution (DE) [28], simulated annealing (SA) [29], gray wolf optimizer (GWO) [30], and artificial bee colony (ABC) algorithm [31], etc. It is beyond the scope of this book to discuss about all the hybridization techniques employed in the SCA algorithm. However, for giving a fair idea to the readers, some of the hybrid algorithms concerning to SCA algorithm are discussed below.

Elaziz et al. [28] proposed a hybridization of sine cosine algorithm (SCA) with differential evolution (DE) algorithm for tackling the feature selection problem. The proposed hybrid algorithm is called SCADE, which has the strengths of DE algorithm and SCA algorithm combined. The feature selection problem is a binary optimization problem, so that solutions in the population represent the binary vectors with length equal to the number of features. Suppose X_i is a solution, and elements of x_i will take values 1 or 0, where 1 represents the selection of the particular feature, while 0 represents the non-selection of the feature. The underlying objective function for evaluating the fitness of the solutions is mentioned below:

$$f(X_i) = \psi \times \text{Err}_{X_i} + (1 - \psi) \times \left(1 - \left(\frac{|S|}{D}\right)\right) \qquad (5.49)$$

where Err_{X_i} represents the classification error of the logistic regression classifier with respect to the solution X_i, $|S|$ is the number of selected features, and D is the total number of features in the given data set. $\psi \in [0, 1]$ is a random number used to balance the accuracy of the classifier and the number of selected features.

The normalized fitness value for each solution X_i, s.t. ($i = 1, 2, \ldots, N$, and N is the total number of features), is computed using the following equation:

$$\text{Fit}_i = \frac{f_i}{\sum_{i=1}^{N} f_i} \qquad (5.50)$$

The best solution (say) P_g is determined from the population after assigning a fitness value to every solution in the population. In updating phase, the position update mechanism of the DE algorithm or SCA algorithm is used, according to a random value $p \in [0, 1]$. Suppose X_i is a solution and Fit_i represents the normalized fitness value of the solution X_i, if $\text{Fit}_i > P$, then position update mechanism of DE algorithm is utilized, else ($\text{Fit}_i \leq P$) position update mechanism of the SCA algorithm is used. The performance of the hybrid SCADE was tested on UCI datasets, and significant improvement in the classification accuracy of the logistic regression classifier is reported [28]. The hybridization technique is an effective technique for improving the performance and robustness of the underlying meta-heuristic algorithm. In the similar fashion, we will discuss below the hybridization of sine cosine algorithm (SCA) with gray wolf optimizer (GWO), and the hybridization of sine cosine algorithm with Particle Swarm Optimizer (PSO).

Singh et al. [30] proposed a hybridization of gray wolf optimizer (GWO) [32] and sine cosine algorithm (SCA) [2]. In the proposed hybrid algorithm, the exploitation phase utilizes the GWO algorithm, while the exploration phase incorporates the exploration capabilities of the SCA algorithm. A randomly generated population is initialized, and the fitness value of each search agent is calculated. Based on the fitness value of the search agents, the best search agent (alpha wolf x_α), the 2nd best search agent (beta wolf X_β), and the 3rd best search agent (delta wolf X_δ) is selected, in the same manner as in the GWO algorithm. After this, for the movement of X_α, the position update mechanism of sine cosine algorithm is utilized. The other parameters involved in the GWO algorithm are kept the same, except the position update mechanism of alpha wolf or the best solution X_α. The position update mechanism of X_α is mentioned in Eq. (5.51).

$$d_\alpha = \begin{cases} r_1 \times \sin(r_2) \times |r_3 \times X_\alpha - X_i| & \text{if } r_4 < 0.5 \\ r_1 \times \cos(r_2) \times |r_3 \times X_\alpha - X_i| & \text{if } r_4 \geq 0.5 \end{cases} \qquad (5.51)$$

$$X_l = X_\alpha - a_l \times d_\alpha \qquad (5.52)$$

where r_1, r_2, r_3, and r_4 are control parameters, as mentioned in the SCA algorithm [2]. And, d_α represents the movement in the X_α. X_l is the next position of the alpha gray wolf [30].

Issa et al. [25] proposed a hybridization of particle swarm optimizer (PSO) with sine cosine algorithm (SCA) in an adaptive manner, and called this hybrid algorithm ASCA-PSO. The proposed hybrid algorithm maintains two layers, namely the bottom layer and the top layer of the solutions based on their fitness value. The bottom layer divides the population into M—different groups, and each group contains N— number of search agents. From every group, a leader y_k (best search agent in a particular group K) is selected for the top layer, and the position update mechanism of PSO is utilized to update the position of the group leaders $y_k, k = 1, 2, \ldots, M$. The bottom layer is responsible for the exploration of the search space and, on the other hand, the top layer is responsible for performing the exploitation phase in the hybrid algorithm [25]. This hybridization ensures a good balance between the exploration and exploitation phase in the entire optimization process [25].

Following the similar trend, Nenavath et al. [33] proposed a hybrid sine cosine algorithm with teaching–learning-based optimization algorithm (SCA–TLBO) to solve global optimization problems and visual tracking. In hybrid SCA–TLBO, first, the standard SCA algorithm is utilized to increase the diversification in the population at the early stages of the search process for exploring the search space extensively, and helping the algorithm in avoiding local optimal regions. After applying the SCA algorithm, search agents are then passed to the teacher-learning phase of the TLBO algorithm in order to move solutions in the direction of the best solution found so far. This strategy helps the proposed algorithm to maintain a fine-tune balance between the exploration and exploitation phase to perform the global and local search effectively.

Gupta et al. [34] proposed the sine cosine artificial bee colony (SCABC) algorithm, which hybridizes the ABC algorithm with the sine cosine algorithm (SCA). The proposed algorithm improves the exploitation and exploration capabilities of the artificial bee colony (ABC) algorithm. In the ABC algorithm, the employed bee phase plays an important role in exploring more promising regions during the search process. The employed bee phase of the ABC is improved using the SCA, it helps the employed bees to prevent irregular exploration, and increases the efficiency of the proposed hybrid algorithm. The position update mechanism of the employed bee phase in the proposed algorithm utilizes the best solution (or elite solution), and is given in Eq. (5.53).

$$
X_i^{t+1} =
\begin{cases}
P_g^t + \left| \frac{f_{best}}{f_{worst}} \right| \times \sin r_2 \times \left| r_3 \times P_g^t - X_i^t \right| & \text{if rand} < 0.5 \\
P_g^t + \left| \frac{f_{best}}{f_{worst}} \right| \times \cos r_2 \times \left| r_3 \times P_g^t - X_i^t \right| & \text{otherwise}
\end{cases}
\tag{5.53}
$$

where P_g^t represents the elite (best) solution at the iteration t. f_{best} and f_{worst} represent the best fitness and worst fitness respectively.

Practice Exercises

1. Discuss the rationale behind opposition-based SCA.
2. Explain the levy flight walk. How this concept is implemented in SCA?
3. Discuss the rationale behind using the covariance concept in SCA.
4. Explain the notion of chaos in Chaotic quantum SCA.
5. What do you mean by encoding? Discuss quantum encoding with suitable examples.
6. How is hybridization going to help SCA in giving better results?

References

1. W. Long et al., Solving high-dimensional global optimization problems using an improved sine cosine algorithm. Expert Syst. Appl. **123**, 108–126 (2019)
2. S. Mirjalili, SCA: a sine cosine algorithm for solving optimization problems. Knowl.-Based Syst. **96**, 120–133 (2016)
3. Y. Shi, R. Eberhart, A modified particle swarm optimizer, in *1998 IEEE International Conference on Evolutionary Computation Proceedings. IEEE World Congress on Computational Intelligence* (Cat. No. 98TH8360) (IEEE, 1998), pp. 69–73
4. M. Suid, M. Tumari, M. Ahmad, A modified sine cosine algorithm for improving wind plant energy production. Indones. J. Electr. Eng. Comput. Sci. **16**(1), 101–106 (2019)

5. N. Kumar et al., Peak power detection of PS solar PV panel by using WPSCO. IET Renew. Power Gener. **11**(4), 480–489 (2017)
6. N. Kumar et al., Single sensor-based MPPT of partially shaded PV system for battery charging by using Cauchy and Gaussian sine cosine optimization. IEEE Trans. Energy Convers. **32**(3), 983–992 (2017)
7. M. Ali, M. Pant, Improving the performance of differential evolution algorithm using Cauchy mutation. Soft Comput. **15**(5), 991–1007 (2011)
8. L.S. Coelho, Novel Gaussian quantum-behaved particle swarm optimiser applied to electromagnetic design. IET Sci. Meas. Technol. **1**(5), 290–294 (2007)
9. X.-S. Yang, *Nature-Inspired Optimization Algorithms* (Academic Press, 2020)
10. A.-F. Attia, R.A. El Sehiemy, H.M. Hasanien, Optimal power flow solution in power systems using a novel sine cosine algorithm. Int. J. Electr. Power Energy Syst. **99**, 331–343 (2018)
11. C. Qu et al., A modified sine cosine algorithm based on neighborhood search and greedy levy mutation. Comput. Intell. Neurosci. **2018** (2018)
12. H.R. Tizhoosh, Opposition-based learning: a new scheme for machine intelligence, in *International Conference on Computational Intelligence for Modelling, Control and Automation and International Conference on Intelligent Agents, Web Technologies and Internet Commerce (CIMCAIAWTIC'06)*, vol. 1 (IEEE, 2005), pp. 695–701
13. M.A. Elaziz, D. Oliva, S. Xiong, An improved opposition-based sine cosine algorithm for global optimization. Expert Syst. Appl. **90**, 484–500 (2017)
14. H. Chen et al., An opposition-based sine cosine approach with local search for parameter estimation of photovoltaic models. Energy Convers. Manag. **195**, 927–942 (2019)
15. J.A. Nelder, R. Mead, A simplex method for function minimization. Comput. J. **7**(4), 308–313 (1965)
16. S. Mahdavi, S. Rahnamayan, K. Deb, Opposition based learning: a literature review. Swarm Evol. Comput. **39**, 1–23 (2018)
17. Y.-W. Jeong et al., A new quantum-inspired binary PSO: application to unit commitment problems for power systems. IEEE Trans. Power Syst. **25**(3), 1486–1495 (2010)
18. M. Soleimanpour-Moghadam, H. Nezamabadi-Pour, M.M. Farsangi, A quantum inspired gravitational search algorithm for numerical function optimization. Inf. Sci. **267**, 83–100 (2014)
19. K. Srikanth et al., Meta-heuristic framework: quantum inspired binary grey wolf optimizer for unit commitment problem. Comput. Electr. Eng. **70**, 243–260 (2018)
20. W. Fu et al., A hybrid fault diagnosis approach for rotating machinery with the fusion of entropy-based feature extraction and SVM optimized by a chaos quantum sine cosine algorithm. Entropy **20**(9), 626 (2018)
21. X.Y. Deng, H.B. Liu, T. Long, A new complex Duffing oscillator used in complex signal detection. Chin. Sci. Bull. **57**(17), 2185–2191 (2012)
22. L. Lv et al., A quaternion's encoding sine cosine algorithm, in *International Conference on Intelligent Computing* (Springer, 2019), pp. 707–718
23. C. Schwartz, Calculus with a quaternionic variable. J. Math. Phys. **50**(1), 013523 (2009)
24. G. Liu et al., Predicting cervical hyperextension injury: a covariance guided sine cosine support vector machine. IEEE Access **8**, 46895–46908 (2020)
25. M. Issa, A.E. Hassanien, D. Oliva, A. Helmi, I. Ziedan, A. Alzohairy, ASCA-PSO: adaptive sine cosine optimization algorithm integrated with particle swarm for pairwise local sequence alignment. Expert Syst. Appl. **99**, 56–70 (2018)
26. H.N. Fakhouri, A. Hudaib, A. Sleit, Hybrid particle swarm optimization with sine cosine algorithm and Nelder-Mead simplex for solving engineering design problems. Arab. J. Sci. Eng. (2019)
27. M.A. El-Shorbagy, M.A. Farag, A.A. Mousa, I.M. El-Desoky, *A Hybridization of Sine Cosine Algorithm with Steady State Genetic Algorithm for Engineering Design Problems* (Springer Nature Switzerland AG, 2020)
28. M.E. Abd Elaziz, A.A. Ewees, D. Oliva, P. Duan, S. Xiong, *A Hybrid Method of Sine Cosine Algorithm and Differential Evolution for Feature Selection* (Springer International Publishing AG, 2017)

29. H. Jouhari, D. Lei, M.A.A. Al-qaness, M. Abd Elaziz, A.A. Ewees, O. Farouk, Sine cosine algorithm to enhance simulated annealing for unrelated parallel machine scheduling with setup times. Mathematics **7**, 1120 (2019)
30. S.B. Singh, N. Singh, A novel hybrid GWO-SCA approach for optimization problems. Eng. Sci. Technol. Int. J. **20**, 1586–1601 (2017)
31. K. Deep, S. Gupta, Hybrid sine cosine artificial bee colony algorithm for global optimization and image segmentation. Neural Comput. Appl. (2019)
32. S. Mirjalili, S.M. Mirjalili, A. Lewis, Grey wolf optimizer. Adv. Eng. Softw. **69**, 46–61 (2014)
33. H. Nenavath, R.K. Jatoth, Hybrid SCA-TLBO: a novel optimization algorithm for global optimization and visual tracking. Neural Comput. Appl. **31**(9), 5497–5526 (2019)
34. S. Gupta, K. Deep, Hybrid sine cosine artificial bee colony algorithm for global optimization and image segmentation. Neural Comput. Appl. **32**(13), 9521–9543 (2020)

Chapter 6
Conclusion and Further Research Directions

The increasing complexity of real-world optimization problems demands fast, robust, and efficient meta-heuristic algorithms. The popularity of these intelligent techniques is gaining popularity day by day among researchers from various disciplines of science and engineering. The sine cosine algorithm is a simple population-based stochastic approach for handling different optimization problems. In this work, we have discussed the basic sine cosine algorithm for continuous optimization problems, the multi-objective sine cosine algorithm for handling multi-objective optimization problems, and the discrete (or, binary) versions of sine cosine algorithm for discrete optimization problems. Sine cosine algorithm (SCA) has reportedly shown competitive results when compared to other meta-heuristic algorithms. The easy implementation and less number of parameters make the SCA algorithm, a recommended choice for performing various optimization tasks. In this present chapter, we have studied different modifications and strategies for the advancement of the sine cosine algorithm. The incorporation of concepts like opposition-based learning, quantum simulation, and hybridization with other meta-heuristic algorithms have increased the efficiency and robustness of the SCA algorithm, and meanwhile, these techniques have also increased the application spectrum of the sine cosine algorithm. The integration of machine learning techniques with the sine cosine algorithm is also becoming a hot topic for researchers in various fields. For instance, Zamli et al. incorporated the concept of Q-learning along with the levy flights in the sine cosine algorithm (QL-SCA) for improving the performance of the algorithm [1]. Wang et al. combined the mechanism of extreme learning machines (ELM) and sine cosine algorithm to develop a novel air quality prediction model [2]. Fu et al. utilized the sine cosine algorithm with variational mode decomposition (VMD) model, and singular value decomposition (SVD)-based phase space reconstruction (PSR) method to optimize least squares support vector machine (LSSVM) [3]. Similarly, Sahlol et al. utilized the sine cosine algorithm (SCA) for the training of multi-layer perceptron (MLP) or feedforward neural networks (FNNs) to improve the prediction of liver enzymes on fish farmed on nano-selenite [4]. The SCA algorithm is used in

© The Author(s) 2023
J. C. Bansal et al., *Sine Cosine Algorithm for Optimization*,
SpringerBriefs in Computational Intelligence,
https://doi.org/10.1007/978-981-19-9722-8_6

the training phase of NN to update the weights and the biases of the network. Hamdan et al. used the sine cosine algorithm (SCA) to train an artificial neural network (ANN) for forecasting the electricity demand [5]. And, Song et al. utilized sine cosine algorithm (SCA) to optimize the classification performance of the back-propagation neural network (BP-NN) for the image classification task [6]. The weights of the BP-NN were optimized using the SCA in the training phase to improve the classification accuracy.

Despite having all the characteristics of a good optimizer, the sine cosine algorithm still needs more attention from the researchers and focused application-oriented approaches in the development. The incorporation of various advance strategies and modifications can enhance the performance and optimization capabilities of the sine cosine algorithm.

References

1. K.Z. Zamli et al., A hybrid Q-learning sine cosine-based strategy for addressing the combinatorial test suite minimization problem. PLoS ONE **13**(5), e0195675 (2018)
2. J. Wang et al., An innovative hybrid model based on outlier detection and correction algorithm and heuristic intelligent optimization algorithm for daily air quality index forecasting. J. Environ. Manage. **255**, 109855 (2020)
3. W. Fu et al., A hybrid approach for measuring the vibrational trend of hydroelectric unit with enhanced multi-scale chaotic series analysis and optimized least squares support vector machine. Trans. Inst. Meas. Control **41**(15), 4436–4449 (2019)
4. A.T. Sahlol, A.A. Ewees, A.M. Hemdan, A.E. Hassanien, Training feedforward neural networks using sine cosine algorithm to improve the prediction of liver enzymes on fish farmed on nano-selenite (IEEE, 2016)
5. S. Hamdan, S. Binkhatim, A. Jarndal, I. Alsyouf, On the performance of artificial neural network with sine cosine algorithm in forecasting electricity load demand, in *International Conference on Electrical and Computing Technologies and Applications (ICECTA)* (2017)
6. H. Song, Z. Ye, C. Wang, L. Yan, Image classification based on BP neural network and sine cosine algorithm, in *IEEE International Conference on Intelligent Data Acquisition and Advanced Computing Systems: Technology and Applications* (2019)

Index

© The Author(s) 2023
J. C. Bansal et al., *Sine Cosine Algorithm for Optimization*,
SpringerBriefs in Computational Intelligence,
https://doi.org/10.1007/978-981-19-9722-8

Printed in the United States
by Baker & Taylor Publisher Services